COMMUNICATIVE BEHAVIOR AND EVOLUTION

Contributors

Robert B. Cairns
Joseph B. Couch
Irenäus Eibl-Eibesfeldt
Roger S. Fouts
John L. Fuller
Benson E. Ginsburg
Martin E. Hahn
Peter H. Klopfer
Thomas K. Pitcairn
J. P. Scott
Edward C. Simmel

COMMUNICATIVE BEHAVIOR AND EVOLUTION

EDITED BY

MARTIN E. HAHN
William Paterson College
Wayne, New Jersey

EDWARD C. SIMMEL
Miami University
Oxford, Ohio

ACADEMIC PRESS New York San Francisco London 1976

A Subsidiary of Harcourt Brace Jovanovich, Publishers

ACADEMIC PRESS, INC.
111 Fifth Avenue, New York, New York 10003

United Kingdom Edition published by
ACADEMIC PRESS, INC. (LONDON) LTD.
24/28 Oval Road, London NW1

Library of Congress Cataloging in Publication Data
Main entry under title:

Communicative behavior and evolution.

Based on papers presented at the Symposium on the Evolution of Communicative Behavior held at William Paterson College in Nov. 1974.
Includes bibliographies and indexes.
1. Interpersonal communication–Congresses. 2. Psychology, Comparative–Congresses. 3. Genetic psychology–Congresses. I. Hahn, Martin E. II. Simmel, Edward C. III. Symposium on the Evolution of Communicative Behavior, William Paterson College, 1974. [DNLM: 1. Communication–Congresses. 2. Social behavior–Congresses. HM258 S989c 1974]
BF637.C45C65 153 75-19644
ISBN 0-12-314350-0

PRINTED IN THE UNITED STATES OF AMERICA

CONTENTS

LIST OF CONTRIBUTORS

Numbers in parentheses indicate the pages on which the authors' contributions begin.

Robert B. Cairns (115), Department of Psychology, University of North Carolina, Chapel Hill, North Carolina

Joseph B. Couch (141), Institute for Primate Studies and Department of Psychology, University of Oklahoma, Norman, Oklahoma

Irenäus Eibl-Eibesfeldt (81), Arbeitsgruppe für Humanethologie am Max-Planck-Institut für Verhaltensphysiologie, Starnberg, West Germany

Roger S. Fouts (141), Institute for Primate Studies and Department of Psychology, University of Oklahoma, Norman, Oklahoma

John L. Fuller (23), Department of Psychology, State University of New York, Binghamton, New York

Benson E. Ginsburg (59), Behavioral Genetics Laboratory and Department of Biobehavioral Sciences, University of Connecticut, Storrs, Connecticut

Martin E. Hahn (1, 163), Department of Biology, William Paterson College, Wayne, New Jersey

Peter H. Klopfer (7), Department of Zoology, Duke University, Durham, North Carolina

Thomas K. Pitcairn (81), Arbeitsgruppe für Humanethologie, Am Max-Planck-Institut für Verhaltensphysiologie, Starnberg, West Germany

J. P. Scott (39), Center for Research on Social Behavior, Bowling Green State University, Bowling Green, Ohio

Edward C. Simmel (1, 163), Behavior Genetics Laboratory and Department of Psychology, Miami University, Oxford, Ohio

PREFACE

The contributors to this volume represent a number of scientific disciplines (zoology, ethology, behavior genetics, comparative psychology, and developmental psychology); they work from a variety of theoretical viewpoints, employing diverse research strategies and using many different species as the subjects of choice. Nevertheless, we think that the chapters taken together present a cohesive whole—that the variety of backgrounds and viewpoints serve to minimize redundancy, and to present as true a picture as possible of the nature and implications of this area of investigation at this time. To this end, we hope that the following brief summary of the chapters to be found in this volume will prove of interest to potential readers.

Klopfer provides the "keynote" chapter, presenting a critical analysis of past and current attempts to deal with the evolution of behavior in general, and of communication and human language in particular.

Fuller specifically introduces what proved to be a major theme of several of the chapters, the behavior-genetic approach. He summarizes many data on communicative behavior taken from a wide variety of species, followed by a succinct view of the problems and opportunties to be found in applying behavior-genetic analyses to determine the sources of intraspecific differences in communication.

Scott carries forward the behavior-genetic theme, providing an original systems approach to genetic and selection mechanisms, then showing how this approach applies to communicative behavior through his current research on dogs.

Ginsburg, while also relating to the behavior-genetic theme previously mentioned, emphasizes the importance of communication in establishing social roles and social organization, supported by original data from his canine research program (dogs, wolves, and coyotes). One of the unique features of this research is the investigation of *interspecific* communication.

Pitcairn and Eibl-Eibesfeldt discuss communication from the perspective of modern ethology, emphasizing behavioral commonalities within a species and providing an interesting contrast with several of the chapters that precede and follow. Most readers will not be surprised by the well-illustrated experiments, but some might be surprised by the species used.

Cairns' approach contrasts directly with the one immediately preceding it. Supported by findings from many sources, including his own data from inbred mice and human infants, he emphasizes the importance that the ontogeny of social interactions has for a theoretical formulation of the evolution of communicative behavior. Important implications of this approach for research strategy are also discussed.

Fouts and Couch present their latest, and perhaps most startling findings on a different sort of interspecific communication: learned language in chimpanzees, together with a review of the work Fouts began with the Gardners. We placed this chapter last partly because we hoped that the reader might leave this book with a question: "What is human communication?"

Although this volume had its origins in the Symposium on the Evolution of Communicative Behavior held at William Paterson College in November, 1974, it is not a transcription of those proceedings, but rather must be considered an outcome of them. Each contributor prepared his chapter following the symposium so that he could take into account the other presentations, as well as the formal and informal discussions.

While held on a New Jersey campus, the symposium is considered the third in the series, Miami University Symposium on Social Behavior, and bears a strong resemblance in organization and conceptualization to its two predecessors and to the volumes which were their outcomes: *Social Facilitation and Imitative Behavior* (edited by E. C. Simmel, R. A. Hoppe, and G. A. Milton, and published by Allyn & Bacon in 1968); and *Early Experiences and the Processes of Socialization* (edited by R. A. Hoppe, G. A. Milton, and E. C. Simmel, and published by Academic Press in 1970).

ACKNOWLEDGMENTS

Preparation of the manuscript for this book was made possible by support from the College of Arts and Science of Miami University. We are indebted to its dean, C. K. Williamson, for his support and encouragement. And, of course, this book would not have appeared so promptly without the expert services of our manuscript typist, Mrs. Betty Marak; our indexer, Sonja Haber; and the excellent work of the editorial and production staffs of Academic Press.

The symposium itself was sponsored by two units of William Paterson College: the Student Government Association and the College of Arts and Science and the dean of the latter, Jay Ludwig. We gratefully acknowledge this support.

The symposium could not have been as successful and enjoyable as it was without the help of Dean Alvin Shinn and Ms. Dorina Frizzera, and of a number of students who gave freely of their time: Charlene Buser, Denise Degethoff, Judy Grapes, Herb Hanson, Diane Hughes, Manny Lamprou, John McGraw and Marie Vitale. Miami graduate students John Wright, Greg Harshfield, and Sonja Haber also provided invaluable assistance, as did Karen Glovinsky.

We are also most grateful to Marilyn and Lynn who helped in many ways.

Edward C. Simmel
Martin E. Hahn

Introduction

DIVERSITY, EVOLUTION, AND COMMUNICATION

Edward C. Simmel
Miami University

Martin E. Hahn
William Paterson College

Communicating to others about communication can be a lot like thinking about thinking—using a process to describe that same process can lead one to think in circles. Therefore, it might be reassuring to the reader to find that this book is really about evolution. More specifically, it concerns natural selection as a major factor in evolution, with emphasis on the role played by social interaction. Since communicative behavior is an integral part of virtually any social interaction process, it can serve as a conveniently measurable, interesting, and perhaps basic area of focus and point of departure.

SOCIAL INTERACTION AND COMMUNICATION

The types of social interactions for which communicative behavior plays the most obvious and clear-cut role in providing specific reproductive advantage are those involved in courtship, sexual behavior, and care of young, as well as instruction (via imitation) in the means of locomotion, defense, and food acquisition. It might seem obvious that in those species where fixed action patterns are found, a great many of these patterns would revolve around sexual and maternal behavior. However, one should not ignore the importance of the greater variation of communicative behaviors concerned with these same aspects of behavior in those species not displaying a very narrow range of species-typical responses.

The role of communication in many other forms of social interactions may be less obvious and direct, but equally important in their evolutionary and adaptive implications. Consider, as just one example, those sorts of behavior communicating threat, dominance, or similar agonistic meaning. Not only do these signals serve to reduce or even prevent actual physical encounters, and thus reduce the likelihood of death or physical injury to the individuals involved, but they also suggest a potentially greater evolutionary significance. Depending on the species involved, and the setting, such displays could maintain the integrity of a group so that its size and structure are optimal for food acquisition, reproduction, and care of young. Or, if shown as an aspect of territoriality, agonistic displays can assist in maintaining a suitable population density. While natural selection can operate on many types of individual behaviors as well as on many nonbehavioral phenotypes, effects on social behaviors, involving as they do more than one animal, can be multiplied accordingly.

Although the papers in this volume deal with various aspects of communicative *behavior,* we should mention at the outset that not all communication need be behavioral (on the part of the sender) in order to result in potentially adaptive responses on the part of a receiver. Consider, for example, the mere presence of a morphological characteristic (a shark fin), and the individually adaptive response elicited from a potential swimmer (stay out of the water). No threat gestures would be offered here, and none are required.

BEHAVIOR AND EVOLUTION

Does an animal behave as it does because only those behaviors that have proved successful in enabling its ancestors to exist in this cold, cruel world survive; those behaviors having been programmed into its nervous system? Or do "higher" species (those with relatively complex nervous systems and low reproduction rates) remain because their behavior is almost infinitely plastic and randomly emitted in a changing environment that rewards with survival those who emit the appropriate responses? These questions express (albeit in rather extreme terms) the two sides to the classical nature–nurture issue. There is no definitive answer, possibly because by now most workers in evolution and behavior recognize that it was not a very good question to begin with.

Naive as the nature versus nurture, instinctive versus learned questions may seem at present, it is useful to remember that the academic ancestors to most research on behavior and evolution were heavily influenced by them. Consequently, it might be helpful to take a very quick look at the two sides to the old controversy, and to suggest what has become of it, granted that it has not been (and probably cannot be) resolved.

While the nature side of the controversy (represented by the early ethologists) differed from the nurture side (represented by the classical behaviorists) in terms of the plasticity of behavior, as well as on some fundamental philosophical and methodological points, it is sometimes forgotten that they both agreed on two fundamental issues: First, and most important, they did agree that the most important "use" of behavior was that of enabling an animal to adapt to (i.e., survive in) its environment. Behavior was seen as being involved in evolution, and as more than mere spin-off from morphological characteristics and physiological or biochemical underpinnings. Second, both camps seemed to assume the lack of variation within a species' most important forms of adaptive behaviors; the ethologists obviously by seeking species-typical behaviors for study; the behaviorists by tight control to reduce "error variance" in quest of "general laws" of behavior.

We do not mean to be unfair by giving polarized and somewhat oversimplified positions to the two original schools in this controversy, and by no means wish to negate the many important contributions made by both classical ethologists and classical behaviorists. However, our concern here is with the understanding of the relationship between behavior and evolution. The extent to which the understanding of this relationship has been advanced beyond the early polarized arguments should be apparent in the chapters that follow.

What has changed, and why? With the original nature–nurture question now silent, or providing at most a soft undertone, new approaches and assumptions have been added to the best of the earlier findings and techniques.

Of obvious importance is the development, over the past several years, of an increasing biological sophistication among students of animal behavior: The animal kingdom is *not* arranged along a "phylogenetic scale" with man at the top and amoebae at the bottom, and the history of behavior is not neatly displayed among "levels" of currently living forms. Furthermore, behaviors are "adaptive" not merely because they aid in the survival of the individual, but rather because they provide a relative reproductive advantage *and* meet various criteria of relationship to the underlying genotype, thus increasing in frequency within a population.

Along with an increase in general biological sophistication has come the rapid development over the past 15 years of the interdisciplinary study of behavior genetics. Behavior-genetic analysis provides a means for determining how behavioral phenotypes (psychophenes) can affect the evolutionary process. An increasing understanding of behavior genetics has helped to take the "versus" out of the nature–nurture controversy: A phenotype that is largely genetically determined is not necessarily immutable in every environmental setting. Behavior-genetic analysis provides methods for determining specific genotype–environment coactions (or more accurately, coactions between environments and phenotypes of known genetic determination). The utility of behavior genetics to the study of evolution and communicative behavior is strongly reflected in many of the chapters in this volume. In a similar fashion, an understanding of the ontogeny of behavior can increase the level of sophistication of evolutionary explanations of behavior. As behavior genetics emphasizes behavioral differences within and between populations, so the ontogenetic approach shows us that behaviors at different stages of development can have varied degrees of impact and evolutionary significance.

EMPHASES AND OMISSIONS

The chapters that follow present a reasonably thorough picture of the current "state of the art" of the investigation of the relationship between evolution and social behavior (exemplified by communicative behavior) as a major part of this investigative effort exists at the present time. The papers represent a variety of approaches: general evolutionary theory, phylogeny, ontogeny, and behavior genetics, as well as combinations of these. Some are concerned with the evolution of communicative behavior, others with communication and social interaction as factors in evolution, and one involves the evolutionary implications of a form of behavior that does not exist in nonhuman primates in their natural setting. Clearly, communicative behavior in mammals is emphasized in this volume, with a large number of species represented: chimpanzees, monkeys, mice, dogs, wolves, and humans—infants, children, and adults.

As with any project of this type, decisions must be made regarding the scope and limitations to be imposed. In order to provide sufficient depth within the topics and approaches covered, and to depict the diversity of currently viable points-of-view, while at the same time producing a volume of manageable size, certain topics were deliberately slighted. While these topics are of considerable importance and potential relevance to the present project, we feel that they are well covered elsewhere and would not be given justice within our space limitations. One topic is pheromonal communication, especially as this is found in the social insects, as well as other aspects of communicative behavior in the social insects. We have not included very much in psycholinguistics, since this has become a highly specialized field of study in its own right.

As mentioned, the original papers that follow show a diversity of viewpoints. We neither decry this fact, nor do we attempt a premature synthesis, but rather believe that as natural selection acts upon diverse phenotypes, so will the evolutionary processes of science select those that prove to have the greatest heuristic value.

Chapter

1

EVOLUTION, BEHAVIOR, AND LANGUAGE[1]

PETER H. KLOPFER
Duke University

EVOLUTION AND CHANGE

The study of evolution has generated metaphors applicable to fields as diverse as biology and economics. Metaphors are usually the currency of poets, which is one of the reasons poetry serves science well (Bateson, 1972). The insights of the poet, however, are no substitute for the statements that serve scientists as explanations: These require explicit referents, a formal structure, a heuristic

[1] My research and the opportunity for such reflections are supported by MH-004453 and Research Scientist Award.

function, and, at their best, an elegant simplicity. Metaphors are out of place in this language. Thus, while it may be appropriate in some contexts to refer to natural selection and the evolution of individuals, or economic systems, or societies, we should recognize these as metaphorical statements. In the currency of Darwin's evolution by natural selection, whether the original mint of the 1840s or a more recent coinage, it is only lineages that evolve.

A lineage is a temporally ordered sequence of individuals derived from common ancestors, whose co-extant members maintain some specified degree of consanguinity. Differences in reproductive success among the ancestors lead to some genotypes becoming disproportionately represented in future generations. As a consequence of selective mortality or differential fecundity, the surviving individuals may thus come to differ from their predecessors in many ways—in their appearance, in their behavior, or in the function of certain organs. Yet, it is not size or shape, the kidney, or maternal behavior that have evolved through natural selection. These have merely changed. It is the lineage that has evolved.

No society can, *sensu strictus,* evolve. The temporally distinct stages of a society are not bound by the genetic ties that hold together a lineage. Neither do these stages correspond in any but a metaphorical sense to the generations that mark the succeeding stages of a lineage.

METAPHORS AND BEHAVIOR

The use of metaphors is particularly common among ethologists, and for good reason: Behavior, the focus of ethological study, is also the focus of studies in religion, advertising, and political sciences. This does not imply any identity in purposes, methodologies, or perceptions, but to the degree that there is any overlap (and the writings of those premier ethologists, K. Lorenz, N. Tinbergen, and W. H. Thorpe clearly show there is much), the metaphorical license allowed divines, Madison Avenue men, and politicians is assumed to be appropriate to ethologists as well.

Behavior is a noun, though that which an animal does is a

process. Behavior, sans italics, is not a noun, a subtle distinction often overlooked. "Stamp out nouns" is the wry motto of mathematician Anatolo Holt (in Bateson, 1972), his way of asserting the dilemma that arises when our manner of speaking influences our manner of thinking (Hardin, 1960; Whorf, 1956).

What is behavior, if it is not a noun that is applicable to some palpable aspect of an organism's state?

A facile, and not altogether misleading response, is that behavior is what ethologists study. Their studies generally focus upon structural or functional aspects. In the latter instance, the questions that are asked may refer to the value to the organism of doing what it does. For example, Hailman (1967) noted that newly hatched gull chicks peck at the tips of one another's bills. Since their bill tips are also marked by a color that contrasts with the remainder of the bill, the pecking response has a distinctly purposeful aspect. Hailman measured the time required by naive, individually isolated gull chicks to discover the edible contents of a can of cat food. He compared this to time required by pairs of chicks. These, in turn, were compared with pairs one member of which had had prior feeding experience. The paired naive chicks took as long to find the food as the solitary birds, though the pecking rates (pecking at spots and specks) of the paired birds were higher. But, in the pairs that included one experienced chick, the naive partner discovered the food speedily. The process seemed obvious: The experienced chick went directly to the food, but because of the mutual acts of bill-tip-pecking the naive chick was also led directly to the food. In short, the value of bill-tip-pecking is that it facilitates recognizing what is edible. This is mutually advantageous due to the circumstance that chicks physically close enough to one another to bill-peck are also (in nature) genetically similar members of the same brood. Hence, according to conventional theory, genetic fitness can be expected to be enhanced by their behaving altruistically and sharing experiences and food.

We may also ask what the history or evolutionary past of this response has been. Did it spring full-blown, an unexpected mutant, or did it develop gradually over succeeding generations? Indeed, it was questions of this sort that particularly excited Konrad Lorenz. To this day he considers the historical portion of his work his major scientific contribution (Lorenz, 1974). His extensive descriptions and speculations on the origins of the courtship displays of the dabbling ducks provide an illustration of this approach as does an equally historically oriented account of courtship in Empeied flies by T. Schneirla (1953; and see Klopfer, 1973).

A structural approach, in contrast, involves looking at physiological mechanisms and ontogenesis. For instance, what mechanism leads the laughing gull chick *(Larus atricilla)* to prefer to peck red spots rather than spots of another hue (Hailman, 1967)? The answer appears to lie in the physical characteristics of oil droplets in the retina. These droplets contain pigments that act as differential filters. When longer (red) wave lengths predominate, neurons that produce a pecking response are excited; with shorter waves, inhibitory processes dominate and pecking declines.

In both approaches, that of the structurally oriented etholgist or that of the functionally oriented ethologist, the description of the process under study is in terms of rates, sequences, and patterns of movement—in short, the object of ethological study, behavior, is a process, and, like all processes, must be described dynamically. There can be no photograph of "a behavior," though partial records of behavior patterns, if fossilized, are a possibility—the sonogram of a bird's song, for instance (Marler, 1963; Nelson, 1973) or a multidimensional scalogram of the golden jackal's *(Canis aureus)* precopulatory display (Golani, 1973). "Behavior" encompasses those activities of animals that ethologists study, and those activities involve changing states in the underlying mechanisms.

HEREDITY AND BEHAVIOR

Those processes of an organism that we identify as behavioral hardly differ except in scale from such other events as those called metabolic or biochemical. The potential for change of these last-named processes is generally accepted because their hereditary bases are at least partially known. Thus, the ability to digest lactose is dependent on the ability to synthesize the enzyme lactase, which, in turn, may be related to a specific sequence of base triplets in the DNA of the chromosomes. The chromosomes and the information their DNA generates are the actual link between generations. This material is, in a literal sense, passed on from parent to offspring. To the degree that a process is the expression of this material it, too, is heritable.

There is some oversimplification in this account, of course. Genes, i.e., a section of the chromosomal DNA, do not directly secrete enzymes such as lactase. Rather, they initiate the synthesis (in response to external events?) of a species of RNA, which reacts with extrachromosomal substances to produce secondary products that may activate or repress other sections of the DNA leading to the production of yet other species of RNA. From cycle to cycle the molecules travel. The separate contributions of the literally inherited material and the material provided by its environment cannot be distinguished. They cannot be distinguished because the process of enzyme synthesis is fundamentally one of interactions. Nonetheless, in a given environment, there is a degree of constancy among the offspring of particular parents, a degree of predictability respecting their properties, and, although improperly, we do speak of the heritability of the capacity to digest lactose.

But what of behavior? To what extent can we speak of those processes which we label "behavior" as being heritable?

Ethologists who have focused on the behavior of fruit flies (*Drosophila* sp.) have perhaps come as close as anyone to an answer. They have been able to compare the behavior of animals whose genotypes differ at only one locus. The effects of single genes upon wingbeat frequency and phototaxis have been noted, for example. These responses in turn may influence courtship behavior and mating behavior (Manning, 1965). It is certainly reasonable in such cases to conclude that the differences in behavior are heritable and even attributal to particular segments of the DNA. Yet, even in such seemingly simple cases a caveat is in order: The surgical removal of a fly's antennae or wings alters its behavior. Various genes can lead to a similar loss. However, the effects of the gene are more far-reaching than the effect of a surgical intervention. Further, the gross changes in reproductive behavior that may be attributed to single-gene differences are gross because of nongenetic feedback. A gene that results in the elimination of red pigment from the eyes of males renders them nearly blind. This not only hampers the males in keeping up with females, but also reduces the excitatory effects of the females upon the males, which, in turn, makes the males less attractive to the females (see Manning, 1965). In short, the existence of gene-dependent, heritable differences in behavior does not allow the conclusion that a behavioral process is itself inherited.

The significance of this interactive character of genes in development becomes even greater when we consider birds and mammals, whose behavioral repertoires are apparently broader than those of insects—and little wonder given the vast increase in the quantity of neural material. The tremendous capabilities vertebrate organisms show for buffering themselves against environmental or genetic onslaughts—maintaining constancy of output in the face of change—as well as their complementary plasticity, argues for behavioral development's being a process that involves multiple feedback controls, rather than the lineal unfolding of a preprogrammed design. What this leads to is a view of behavior as a system that has multiple controls, including some that are genetic. The pathways of causality and control are numerous, however, with many cross-overs, so it is rarely, if ever, possible to isolate single heritable factors whose effects are inevitable and irreversible.

EXTRAPOLATION AND BEHAVIOR

The foregoing has a profound relevance to the issue of behavioral evolution and the attempts to understand one species by studying another. Moral constraints prevent our rearing human infants on wire or terrycloth surrogates, but these constraints seem not to apply to young monkeys. Economic considerations may lead to the substitution of rats for Rhesus, and I know of one psychology laboratory where a computer program replaced the rats (that was cheaper yet). But, in all these cases there is implicit the belief that whatever is learned from one organism will be applicable to an understanding of another.

The intellectual basis for this belief in the continuity of mental ability was provided by Darwin's theory of evolution and reinforced by the synthesis between Darwinism and modern genetics. Even before Darwin, however, comparative studies of morphology by Owen and others had made distinctions between similarities due to common function *(analogues)* and those due to structural correspondence *(homologues)* (cf. Campbell & Hodos, 1970). After Darwin, structural correspondence or homology was attributed to common descent. The distinction between homology and analogy is crucial for it separates similarities due to kinship (genetic similarity), which are the similarities vital to extrapolation, from similarities due to accident or convergent evolutionary pressures. Pineapples and alligators have a skin that seems similar but it would rarely be taken as a sign of kinship! Other cases of convergence may be less readily distinguished from relationship.

The point is this: Extrapolation from one species to another depends upon distinguishing those features that are due to a common function (analogues) from those that share a common origin (homologues). Thus, the distinction between analogy and homology must be clear and complete, the classification of actions unambiguous and accurate, if one is to be able to make inferences as to, e.g., the effects of maternal deprivation upon human infants from studies of young apes.

Some ethologists seem confident of their ability to make this distinction. Lorenz (1974) writes:

> As a pupil of the comparative anatomist and embryologist Ferdinand Hochstetter, I had the benefit of a very thorough instruction in the methological [sic] procedure of distinguishing similarities caused by common descent from those due to parallel adaptation. In fact, the making of this distinction forms a great part of the comparative evolutionist's daily work. Perhaps I should mention here that this procedure has led me to the discovery which I personally consider to be my own most important contribution to science. Knowing animal behavior as I did, and being instructed in the methods of phylogenetic comparison as I was, I could not fail to discover that the very same methods of comparison, the same concepts of analogy and homology, are as applicable to characters of behavior as they are in those of morphology. This discovery is implicitly contained in the works of

> Charles Otis Whitman and of Oskar Heinroth; it is only its explicit formulation and the realization of its far-reaching inferences to which I can lay claim. A great part of my life's work has consisted in tracing the phylogeny of behavior by disentangling the effects of homology and of parallel evolution. Full recognition of the fact that behavior patterns can be hereditary and species-specific to the point of being homologizable was impeded by resistance from certain schools of thought, and my extensive paper on homologous motor patterns in Anatidae was necessary to make my point [p. 231].[2]

The assertion itself, however, is polemical, not convincing, and the resistance to which Lorenz alludes is hardly proof that his self-confidence is warranted.

Consider how Lorenz's (1974) argument actually proceeds: A homology is any resemblance between two species that can be explained "by their common descent from an ancestor possessing the character in which they are similar to each other [p. 230]." He then illustrates the structure of the forelimbs of a bat, whale, sea lion, mole, dog, bear, elephant, and man, which, he asserts, serve dissimilar functions but are built upon the same plan (see Figure 1.1). "The very dissimilarity of their functions makes it extremely improbable that the manifold resemblances of their forms could be due to parallel adaptation—in other words to analogy [p. 231]."

Here are the problems: Can you, in light of Lorenz's definition, decide whether a resemblance implies common descent, if the fact of common descent is necessary to define the resemblance? Can you justify operationally or by some other objective means that the forelimbs of various tetrapod vertebrates are indeed dissimilar in function? In my judgment, they are similar, since all serve a locomotory function. Nor do I find the assertion of a similarity in structure compelling in the absence of clear criteria for the determination of degrees of similarity. This definitional problem applies as well to the structures being compared. "Strictly speaking," writes Lorenz, "the term homologous can only be applied to characters and

[2] This and subsequent quotations cited as Lorenz, 1974 are from K. Z. Lorenz, Analogy as a source of knowledge. *Science, 185* (19 July 1974), 229–234. Copyright 1974 by the American Association for the Advancement of Science.

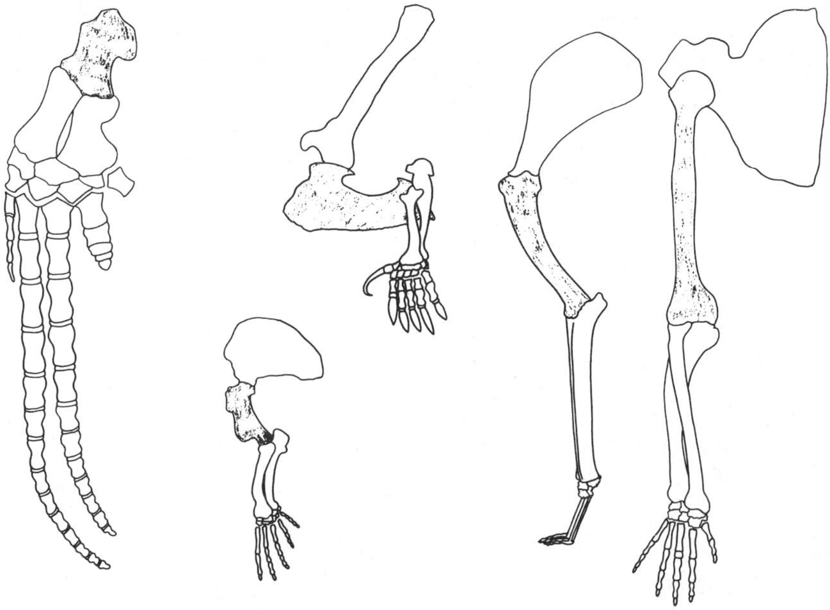

Figure 1.1 Homologies of the vertebrate limb. [Adapted from K. Z. Lorenz, Analogy as a source of knowledge. *Science, 185* (19 July 1974), 229–234. Copyright 1974 by the American Association for the Advancement of Science.]

not to organs [p. 230]." But, how are these two classes differentiated? If one compares light-sensors of an earthworm and a man, one would be inclined to consider them analogous, while a comparison of the constituent cells or their biochemical transductions would lead to their classification as homologies.

In the absence of explicit criteria the "graininess" of the investigator's perceptual film will largely determine his classification. How, then, can the concepts of homology and analogy help us to understand the evolutionary paths of behavior and permit extrapolations from mice to men? Neither these concepts nor the putatively heritable nature of "behavior" appear to me to be useful.

These objections have not, of course, inhibited ethologists and others from drawing conclusions which, if they are true, are certainly not true for the reasons stated. Lorenz has "explained" the evolution of styles in neckwear, railway carriages, love, jealousy, and warfare in man, all by reference to "homologous" structures (see Figure 1.2). This continues to be a popular and accepted sport.

BEHAVIOR AND LANGUAGE

Of the uniquely human attributes whose evolution has been a puzzle, none has intrigued ethologists more than language and its biological roots. The studies of the evolution of human speech thus provide an instructive example of the application both of the classical morphological concepts, homology and analogy, and of an altogether different approach.

In the usual interpretation of evolutionary theory, language is seen as the end-point in a continuum of communicative systems.

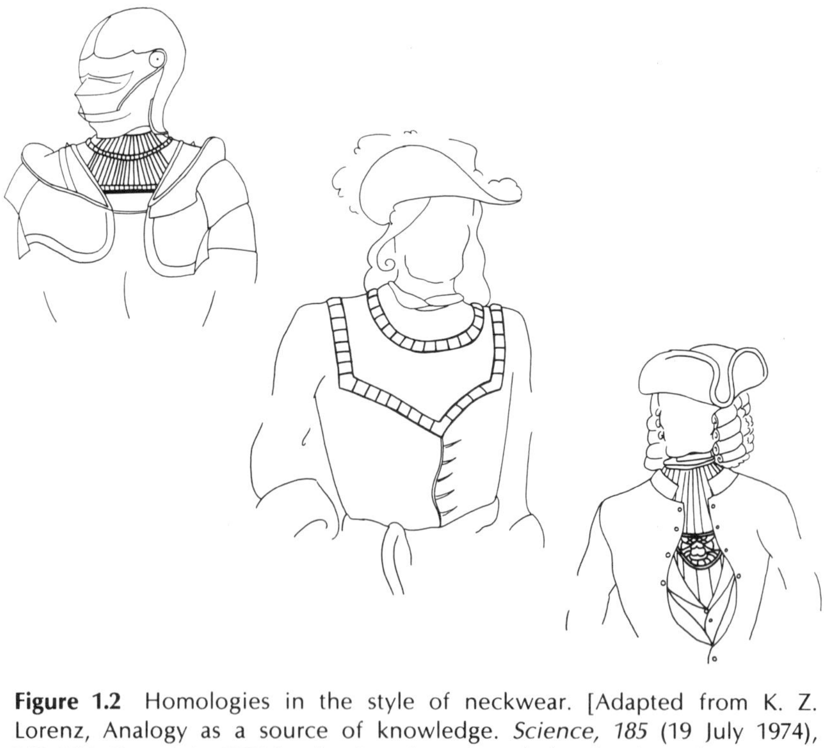

Figure 1.2 Homologies in the style of neckwear. [Adapted from K. Z. Lorenz, Analogy as a source of knowledge. *Science, 185* (19 July 1974), 229–234. Copyright 1974 by the American Association for the Advancement of Science.]

Communication of one sort or another is ubiquitous among social species. There are differences, of course, in the nature of the information that needs to be communicated. In bees, it is the source, distance, and direction of a source of pollen or nectar. Red deer, on the other hand, scarcely require one another for such information. Their communications deal with social status, mood, or other intangibles. There are differences, too, in the sensory systems by means of which communication is achieved. Bees use ritualized dances, scents, and sounds. Deer employ scents, sounds, and particular postures. Some fish depend on neurally controlled color changes or sounds or electric pulses (see Sebeok, 1968, and in press). Amid all of this diversity, however, investigators have nonetheless found common principles that have allowed communicative behavior to be treated in an unitary fashion. This does not deny the existence of discontinuities between phyla or species. Hockett and Altman (1968), for example, have listed some defining attributes of human language. Certain of the elements they list are shared with one other group or species, others with another. The sequence from brute to man, though continuous, has involved the accretion of distinct functions. For example, human language embodies the characteristics of rapid fading, interchangeability of transmitter and receiver, feedback, semanticity, arbitrariness, discreteness, and openness (among others). Bird calls share all of these but the last. In contrast, the postural signs of jackals may preclude interchangeability but not openness.

Whether the evolution of human language is seen as smoothly continuous or as the continual addition of small quantum jumps, it does seem to allow the conventional application of the concepts of homology and analogy.

A contrary point of view has seriously been advanced from time to time, a view that disputes the assumption of evolutionary continuity. Among morphologists and geneticists of note, the most famous recent advocates of saltatory evolution—change through the sudden appearance of "hopeful monsters"—are Petrunkewitsch (1952) and Goldschmidt (1955). Their views are reflected in a study of language origins by Lenneberg (1967), who suggests a *dis*continuity model. The model Lenneberg proposes takes note of two problems continuity theories ignore. First, the problem of identifying evolutionary antecedents. For example, it is entirely possible that the antecedents to human spoken language are not the vocalizations of other animals, but the processes involved in motor coordination or the neurophysiology of perception. Second, the common occurrence of drastic allometric changes in development and, by implication, in evolution produces morphologic discontinuities that are far more readily assumed by the apologists for saltation than those for gradualism. Apparently the seminal work by D'Arcy Thompson on allometric changes in growth, development, and evolution has been lost in the transition to "modern" biology.

Lenneberg's scheme is illustrated by a simple diagram (see Figure 1.3). This is a rather more formal statement of a thesis well known to ecologists, one that we might term "the rules of the game." The thesis is that economic or energetic considerations can shape mature structures so as to obscure similarities at the deep level, or, conversely, that structures that are fundamentally dissimilar may converge as they mature, suggesting a deep relationship that does

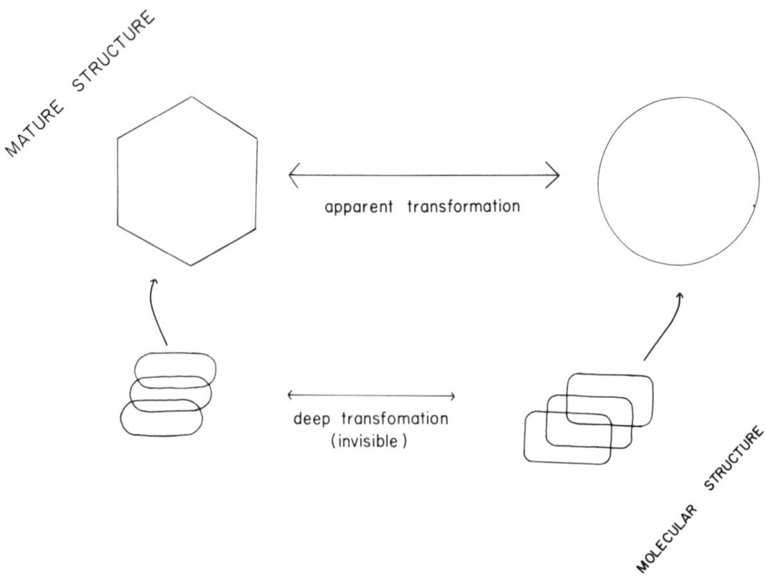

Figure 1.3 An illustration of the relationship between molecular ("deep") structures (homologies) and between mature ("surface") structures (analogies). [Adapted from E. Lenneberg, *Biological foundations of language*, copyright 1967 by John Wiley & Sons, Inc.]

not in fact exist. Defense of a territory, for example, is often taken as an instance of homologous behavior whether it occurs in bird or man. However, the ecological viewpoint leads to a consideration of the costs of territory defense: In terms of the increased exposure to predators, loss of time from other activities (such as feeding), and the exertion of patrol and advertisement, the benefits derived from the exclusive use of a piece of turf must be greater than the costs. It is possible to establish the values of some of these parameters. Thus, it may be shown that where food resources are unevenly distributed in large and randomly moving clusters (as a school of fish), there are unlikely to be benefits from maintaining a feeding territory. Sea birds, in fact, rarely defend areas beyond the immediate confines of their nests, if that. Where food is available in more uniformly spread packets, feeding territories may make economic sense, but only if predation is not an important factor in the species life history (Klopfer, 1973).

According to this view, communicative behavior represents a mature structure, an economically sound response to needs generated by a certain degree of social organization. However, similarities at the level of mature structures provide no information as to the deep relationships.

Presumably, the biological basis for the language potential of human beings does lie at the level of the deep transformations. Hence, we need neither expect similarities at the level of mature structures, nor, finding them, have we a right to assume the similarities to be of consequence.

One set of related molecular structures from which the mature communicative structures could have arisen is the need for contact among fellow creatures, an old view revitalized and championed by S. Langer. Human language itself arose from other forms of contact—tactile, or from the evocative and symbolic powers of dance.

> The only extreme improbability seems to me to be that language arose from some kind of previous communication by improvements that had survival value. Animal contact is not communication: animals may perform acts, even, pick up an act one from another at some juncture, as bees seem to pick up the food-getting act in a sound, without asking or telling anything. It is human mentality that does not remain in the animal pattern—the symbolic finishing of excessive nervous impulses

> . . . begat the first processes of ideation, which eventuate in wild expressions, dance, magic, then the wishing of curses and blessings on other creatures and investing implements . . . with potency, and hallowing the places for dancing or feasting. Speech was born, I believe, in such high reaches of proto-human activity and gathered form when one individual knew by the symbolic utterance of another what that other was thinking. For with such concentrated expression came real envisagement, the beginning of reflection, thought [Langer, 1973, p. 314].

In short, gestures and utterances may elicit ideas held in common from past actions: This is the beginning of communication in the human sense. Language is essentially a system of feeling. Through the possibility it provides of communicating ideas, the moral and mental traits peculiar to man and human society could arise. Faust was wrong. In the beginning was the touch. Then came the word, thought followed. Perhaps we are misled in seeking for the precursors of human language in chimpanzees who use human sign language or manipulate plastic tokens. The common deep structure for which we search may be represented by the rain dance Jane Goodall's wild chimpanzees spontaneously perform. The evolutionary relation between behavior patterns of different animals may yet be perversely obvious if we avoid seduction by the traditional views of comparative morphology.

ACKNOWLEDGMENTS

Special thanks go to D. Rubenstein and L. Reinherz for their criticisms of a draft of this article.

REFERENCES

Bateson, C. 1972. *Our own metaphor.* New York: Alfred A. Knopf.

Campbell, C. B. G., & Hodos, W. 1970. The concept of homology and the evolution of the nervous system. *Brain, Behavior, and Evolution, 3,* 353–367.

Golani, I. 1973. Nonmetric analysis of behavioural interaction sequences in captive jackals. *Behaviour, 44,* 89–112.

Goldschmidt, R. B. 1955. *Theoretical genetics.* Berkeley, California: Univ. of California Press.

Hailman, J. P. 1967. The ontogeny of an instinct. *Behaviour Supplement, 15.*

Hardin, G. 1960. Meaninglessness of the word protoplasm. *Science Monthly, 82,* 112–120.

Hockett, C. F., & Altman, S. 1968. A note on design features. In T. Sebeok (Ed.), *Animal communication.* Bloomington, Indiana: Univ. of Indiana Press. Pp. 61–72.

Klopfer, P. H. 1973. Does behavior evolve? *Annals of the New York Academy of Sciences, 223,* 113–119.

Langer, S. K. 1973. *Mind. An essay on human feeling.* (2nd ed.) Baltimore, Maryland: Johns Hopkins Univ. Press.

Lenneberg, E. 1967. *Biological foundations of language.* New York: Wiley.

Lorenz, K. Z. 1974. Analogy as a source of knowledge. *Science, 185,* 229–234.

Manning, A. 1965. *Drosophila* and the evolution of behaviour. In J. D. Carthy & C. L. Duddington (Eds.), *Viewpoints in Biology 4.* London: Butterworths. Pp. 125–169.

Marler, P. 1963. Inheritance and learning in the development of animal vocalizations. In R. Busnel (Ed.), *Acoustic behavior of animals.* New York: Elsevier. Pp. 228–243; 794–797.

Nelson, K. 1973. Does the holistic study of behavior have a future? In P. P. G. Bateson & P. H. Klopfer (Eds.), *Perspectives in ethology.* New York: Plenum Press. Pp. 281–328.

Petrunkewitsch, A. 1952. Macroevolution and the fossil record of Arachnida. *American Scientist, 40,* 99–122.

Schneirla, T. C. 1953. Basic problems in the nature of insect behavior. In K. D. Roeder (Ed.), *Insect physiology.* New York: Wiley.

Sebeok, T. 1968. *Animal communication.* Bloomington, Indiana: Univ. of Indiana Press. (Second edition, in press.)

Whorf, B. L. 1956. *Language, thought and reality.* New York: Wiley.

Chapter

2

GENETICS AND COMMUNICATION

JOHN L. FULLER
State University of New York at Binghamton

When the editors of this volume invited me to contribute a paper on genetics and communication I was distinctly pleased. For nearly 30 years I have been active in behavior genetics and have argued that behavior–genetic analysis (Hirsch, 1967) is a productive technique that can be applied to a host of problems in biology and psychology. Such analyses can contribute both to understanding the sources of individual differences within populations and to knowledge of the processes through which a minute fertilized ovum becomes a complex organism engaging in multiple transactions with its environment. Communication is certainly one of the most important activities of man and beast; genetics should have something to say about it.

My task seemed to have been already well begun. Geneticists have a vocabulary that implies a communicative role for genes. We speak of genetic information as being expressed in a *language* in which the four nitrogenous bases of DNA—adenine, guanine, thymine, and cytosine—are the four *letters*. Triplet combinations of these letters form either *words,* each of which specifies an amino acid, or a *punctuation point,* indicating the end of a message. Strings of such words, cistrons *(sentences),* provide the information necessary to guide the synthesis of polypeptides, which may combine to form complex proteins *(paragraphs)*. The language analogy has been extended to the processes involved. DNA messages are *replicated* as identical copies. They are *transcribed* to RNA using a similar spelling with slightly different alphabet and are finally *translated* as polypeptide chains.

In multicellular organisms the transfer of information becomes much more complex. Although all vertebrate cells seem to contain the same DNA, only a fraction of its informational content is expressed in any one type of cell. Ginsburg (this volume) contrasts the encoded and the effective genotype. One could extend the metaphorical language to include the whole range of development from conception to death by using such terms as *filing, retrieving, editing,* and *censoring* of DNA information.

Unless analogies are reasonably precise, however, they are misleading rather than helpful. It is intriguing to consider the transfer of genetic information by genes as similar to communication between organisms, but on close inspection the analogy is forced. It rests upon the denial of any distinction between *communication* and *information transfer,* a distinction that is important in the context of this volume. The difference can be illustrated by reference to a newspaper, which certainly provides one with information. Is it a communicator? No. We may apply this term to the reporters, editorial writers, and advertisers who are trying to educate or cajole the readers, but the newspaper itself is only an agent. Could we not, then, consider genes, which are agents for the transfer of genetic information between generations, as the analogues of newspapers? Again, I think not. Except for man, and even here it is probably infrequent, no species mates because it perceives that the act of mating will result in a more or less accurate copy of itself.

As McKay (1972) has emphasized, we destroy the usefulness of the term *communication* if we equate it with information transfer. By simply looking at another person I gain information about age, sex, pigmentation, and possibly many other characteristics. My gain in information can occur even if the other person is unaware of my observing. Something else is required to merit classification as communicative behavior. Information transfer is an essential component, but it is not a sufficient attribute.

It may seem at this point that I am arguing that there is nothing to say about genes and communication, since I have excluded simple information transfer from my class of communicative behavior. Having established this point, I might terminate this essay with a few words of thanks for inviting me to participate. But to do so would be to overlook the genetic component of all behavior. We can inquire about the genetic contribution to observed variation in communicative behavior; we can try to determine the restrictions which genotypes impose upon an individual's ability to acquire species-typical visual and auditory communicative action patterns. We shall look at samples of such researches after a brief digression into the formal properties of communication.

SOME FORMAL PROPERTIES OF COMMUNICATION

The model of communication that I shall employ is based primarily on that of McKay (1972). Information from the environment is collected by receptors, filtered and organized, and passed in coded form to a comparator mechanism. Here the input is matched against information available concerning the current goals of the organism. If the two sources of information match, nothing much happens; if they do not an output is generated. Somehow the output is organized to select from the organism's effectors the actions that tend to change the environment so that the mismatch in the comparator system is reduced. When the surroundings include another individual the output may be a threat display, a courtship dance, an alarm call, or other form of communication. Goals, of course, are transitory and can change quickly. Thus McKay's model includes a hierarchically arranged set of metacomparators and metaorganizing systems. The goal of mating may be set aside if a predator appears and a courtship dance is replaced by an alarm call.

Applying this model to the sender's role in communication we have: (1) perception of another organism whose activities are discordant with respect to the sender's goals; (2) a decision regarding priorities of possible incompatible goals; (3) some type of communicative behavior. The model makes no statement regarding the receiver, and there is no assurance that the message will elicit behavior that furthers the objectives of the sender. Nevertheless, observation of social behavior demonstrates that displays and calls commonly result in a series of communicative acts in which the roles of sender and receiver either alternate between the participants or are maintained simultaneously in each one. The origin of such coordinated patterns is one of the problems for evolutionary theorists.

Strictly followed, McKay's model requires that there exists in the sender's information-processing system an evaluation of the effects of its signals upon other organisms. Since the comparator is not physically specified and is inaccessible to direct observation, this criterion makes it nearly impossible to prove that any animal communicates. Perhaps a looser definition is more useful—*communication is a form of social behavior characterized by the production of relatively stereotyped actions that characteristically modify the behavior of other individuals and elicit relatively stereotyped responses.* Such a criterion is verifiable by an external observer.

Clearly, messages are useful to the sender only when the receiver responds predictably. How are "correct" responses acquired—by experience or from genes? Are the majority of responses conditioned or unconditioned? Probably both types of causes operate synergistically in development at all times, but their relative importance in producing variation differs from case to case. There would be a selective advantage to both participants in having a built-in response to high-priority signals such as alarm calls. Calls employed for individual recognition must certainly be learned by the specific individuals concerned. Here selection would favor a mechanism permitting a degree of plasticity in response to a message; the parent gull's call elicits gaping; a stranger's call leads to aversive action. The rigidity of genetic control of development is itself subject to natural selection.

HOMOLOGY, EPIGENESIS, TABULAE RASA

Every amateur bird-watcher soon learns that the mode of flight, perching, feeding, and singing are as characteristic of a species as its plumage and are sometimes more easily observed. Lorenz (1958) made careful observations on the courtship behavior of various species of ducks, and found that the degree of similarity between species was highly correlated with the closeness of relationship as hypothesized from physical characteristics. Tinbergen (1960) found a similar pattern in his comparative studies of gull behavior. Both sets of observations have been interpreted as evidence for homology of behaviors which are today very different in form. The application of the concept of homology to behavior has been challenged by Klopfer (this volume), but it is not crucial to my topic of genetics and communication. Behavioral similarities are surely less convincing evidence for common ancestry than are basic body plan or DNA structure.

A question of greater interest to a behavior geneticist relates to the genetic control of development. Are the details of visual displays and vocalizations laid down irrevocably in the genotype or are they acquired during development? We know, of course, that the organism is not preformed in the zygote and that its characteristics appear predictably over time provided the environment is favorable. This is the epigenetic point of view which is held universally by present-day biologists and psychologists. But there are rather marked differences of opinion possible within the epigenetic framework. Gottlieb (1971) distinguishes deterministic and probabilistic epigenesis. In the former, genes precisely determine structure which in turn determines physiological function and behavior. Probabilistic epigenesis involves continuous feedback so that function modifies the development of structure and behavior produces changes in function. Behavior geneticists will be found predominantly in the probabilistic camp, but even within this group one will find differences in the relative importance given to the "push" of the genotype and the "feedback" from the environment.

Moltz (1965), for example, writes of the environment as being "actively implicated in determining the very structure and organization of each response system," and contrasts this view with that of a passive, hopefully benign, environment from which an organism takes what it needs. My own bias is toward placing the active role of environment in the phylogeny of a species rather than, as Moltz does, in the ontogeny of individuals. The result is the preprogramming of organisms for survival and propagation in their most probable environments. It is more efficient in the long run for environment to shape the gene pool which codes for successful phenotypes than to start over again in each generation to fit organisms to the available niches.

I shall discuss the varieties of epigenetic thought by an updating of John Locke's (1690) concept of the *tabula rasa* (blank slate). Consider a robot whose actions are programmed by a multichannel tape running through a built-in recorder (see Figure 2.1). As the robot comes off the assembly line (is hatched or born), it is equipped with surface transducers responsive to sound, light, temperature, etc., which can be recorded on appropriate channels. Fed into an integrator, the recorded material determines the motoric output—the robot's behavior. Feedback from previously recorded material to the input (perceptual functions) and to the output provide that past experience as well as immediate stimuli affect behavior. Provided the tape is blank and the input devices nonselective, we have a perfect epigenetic system in which the environment plays the active role in determining the repertory of the robot. But suppose that the tape is imperfect. It may be grainy and incapable of retaining the fine details from the transducers. Some of its channels may have prerecorded

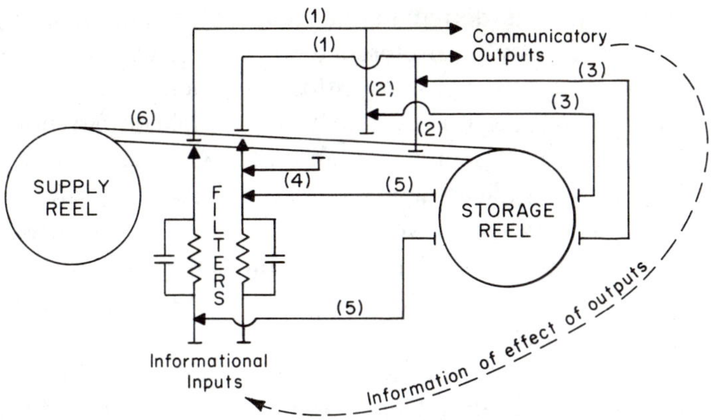

Figure 2.1 Schematic diagram of a tape control system for a communicating robot. Arrows denote direction of transfer of information in the system. ⊥ indicates points of uptake of information. Content of communication depends on (1) current inputs; (2) just prior inputs; and (3) stored inputs. Information inputs are affected by (4) feedback from recent events, and (5) past events. Some channels may carry prerecorded information (6).

material which will affect the robot's behavior, or at least combine with any input to produce a new pattern. It is also possible to insert filters on the input devices so that the patterning of "stimuli" is critical to their recording on the tape. The following section will demonstrate strong evidence that the blank tape model is inadequate to explain the facts of the development of communicative behavior in individuals.

EXPERIMENTAL STUDIES OF DEVELOPMENT

Gottlieb in 1971 reported on a series of studies with Peking ducks, mallards, wood ducks, and domestic chicks which extended over 7 years. The results of this impressive series of experiments are far too extensive to report here, but I shall mention some findings particularly pertinent to genetics. Naive ducklings and chicks, given a choice between recorded maternal calls of their own and other species, showed a strong preference for their own species. Sound stimuli were more effective in eliciting approach than were stuffed birds and other models; however an audio–visual combination was a somewhat stronger attractant than an auditory stimulus alone. Such data might seem to indicate a genetically programmed preference for responding to a mother's vocalizations. The case seemed even stronger after Gottlieb noted that duck embryos reared in incubators could respond to maternal calls 5 days prior to hatching, and that the response was species-specific.

Then eggs were incubated in sound-isolated chambers instead of a communal incubator. Deprived of the stimulation from prehatching peeps of other embryos, these isolate-reared individuals were retarded in development of the specific response to maternal calls. As a final experiment, ducklings were devocalized so that self-stimulation was eliminated as a possible catalyst for development. Such ducklings were even less efficient discriminators between similar calls (chicken versus mallard), and were retarded in ability to discriminate between mallard and wood duck maternal calls. These data demonstrate a strong genetic component that makes Peking ducklings respond selectively to their species call. At the same time, exposure to sound from other ducklings or even self-stimulation speeds up and sharpens discriminatory ability.

Gottlieb's researches were primarily concerned with the specificity of the receiver's reactions. There has been more work on the role of experience in the acquisition of complex communicatory behavior such as bird songs (Thorpe, 1972). Here the multiplicity of findings is surprising. In a number of species the typical song or call develops normally in the absence of a model and even in the presence of an alien song. Included are such diverse forms as the song sparrow *(Melospiza melodia)* (Marler, 1967), and many doves (Lade & Thorpe, 1964). Even early deafening does not hinder normal vocal production of doves (Nottebohm, 1970) or domestic fowl (Konishi, 1963).

In other birds there is a sensitive period during which a relatively specific auditory stimulation is required for species-characteristic sound production later in life. In the chaffinch *(Fringilla coelebs)* this period is in the first autumn some months before singing begins (Thorpe, 1958). In the white-crowned sparrow the sensitive period for exposure is earlier, at a post-hatching age between 10 and 50 days (Marler, 1970). The young males do not immediately imitate their models nor do they imitate the songs of alien species. Instead, exposure seems to initiate the establishment of an "auditory template" which persists in latent form until the bird reaches sexual maturity, when the template guides the formation of a motor pattern, the song. Deafened birds, exposed to their species song at the sensitive period, do not sing properly, though deafening an adult male does not impair his song (Nottebohm, 1970). For our present purpose the most interesting aspect of these data is the rejection of alien songs and the acceptance of the conspecific song. Birds, like our imaginary robot, must either have some preprogrammed complex behaviors or preprogrammed selective filters which pass only specific stimulus configurations.

There is another side of the story. While the calls of laughing gulls may seem to us to be a cacophony of rather unpleasant sounds, these calls appear to play a major part in recognizing individuals within the species (Beer, 1970; Thorpe, 1972). Here the genetic problem was to program development for maximal individualization of vocalizing while retaining species-specific features. To achieve this in our robot we would set the characteristics of the filter within broad limits and arrange it to be precisely tuned to the pattern of the first stimulus received.

COMMUNICATION IN HYBRIDS

The studies just mentioned were concerned with differences between species that are, in a broad sense, genetic. However, a more detailed analysis of the genetic architecture underlying phenotypic variation requires crossbreeding and observations on the phenotypes of the offspring. There are relatively few species that produce fertile hybrids; at least, very few have been studied by scientists interested in communication. However, these cases do permit some analysis of the genetic basis of variation in signals serving similar functions in the two parental species.

One of the early studies was reported by von Hörmann-Heck (1957), who observed mating behavior in two crickets, *Gryllus bimaculatus, G. campestris,* and their hybrids. Three patterns of behavior clearly separate these sympatric species. *G. campestris* vibrates its antennae after pairing; *G. bimaculatus* does not. The mode of inheritance was reported as monogenic with the vibration behavior dominant. *G. campestris* makes pendular movements of the anterior part of the body during copulation; *G. bimaculatus* does not. This characteristic seems to vary continuously as though it were inherited in polygenic fashion with little or no dominance. Finally, *G. bimaculatus* produces precourtship chirps by raising and lowering the elytrae; in *G. campestris* the movements are silent. The segregation ratios suggest monogenic inheritance with heterozygotes intermediate to the parental strains, presumed homozygotes. The three patterns seemed to be inherited independently.

A much more elaborate genetic investigation of cricket songs has been carried out by Bentley and Hoy (1974). The courtship music of the crickets *Teleogryllus oceanicus* and *T. commodus* starts with a chirp followed by a trill. In *T. oceanicus* the chirp is clearly separated from the trill, which consists of a long series of paired sound pulses. In *T. commodus* the end of the chirp is fused with the beginning of the trill, which is shorter with more closely spaced pulses than is characteristic of *T. oceanicus.* There is a striking difference between the reciprocal F_1s that is interpreted as X-linked transmission of some of the genes involved in control of the pattern. Backcrosses tend to be intermediate between the F_1 and the parent species. There is no indication of segregation in the backcrosses for any of 18 quantitative features of the songs. Bentley and Hoy conclude that a large number of independent genes are concerned with laying down the neural network that specifies the temporal aspects of song production.

Females of these species do not sing but their preferences for recorded songs of the males can be measured by means of a Y-maze, which a suspended female holds between her legs. As she walks toward a sound stimulus, the maze moves under her while she remains safely restrained. When given a choice, F_1 hybrid females respond positively to the songs of F_1 hybrid males rather than to either of the parental species songs. Apparently the genes which control the pattern of pulse production are also active in females, though they here manifest their presence by inducing an internalized template. Only songs that match this template elicit approach movements.

Display Patterns in Fish Hybrids

Similar hybridization experiments were conducted with the platyfish, *Xiphiphorus maculatus*, and the swordtail, *X. helleri*, by Clark, Aronson, and Gordon (1954). These species, well-known to aquarium fanciers, do not hybridize in nature but do so in the laboratory. Two distinctive alternative patterns are seen in the precourtship behavior of these fish. Platyfish males peck the sand on the bottom of the aquaria; the more forward swordtails nibble at the females. In the hybrids courtship was generally less vigorous and many of the males failed to show either behavior. The F_1 and F_2 were intermediate; the two backcrosses were closer to the parental species. In another display pattern, swordtails swim backward to touch females with the tail tip (backing) and platyfish approach closely, only to suddenly back away with limp fins (retiring). Here too, the hybrids were intermediate with no clear Mendelian pattern of inheritance demonstrable.

The hybrids were described as showing segments of the courtship behavior characteristic of both parents, but assembled into an uncoordinated total system. And this mismatching was apparent to platyfish females used to test mating success. Mating success in the backcrosses was only 9% and 0% in comparison to 86% and 39% for platyfish and swordtail males, respectively.

Display Patterns in Ducks

A third example of the disruption of display behavior in hybrids is provided by Sharpe and Johnsgard's (1967) observations on a mixed flock of mallards *(Anas p. platyrhyncos)* and pintails *(A. a. acuta)*. These species, though sympatric, seldom hybridize in nature, although the hybrids are fertile. Forty-one F_2, of which 28 were males, were reared in 1962–1963. Those birds who survived to adulthood were evaluated for resemblance to mallard or pintail on two scales, one based upon plumage characteristics, the other upon courtship behavior.

With respect to plumage the F_2 ranged from almost perfect mallards to almost perfect pintails. On a rating scale that gave a score of 0 to mallard and 20 to pintail, two birds scored between 0 and 4, 3 were in the 15–19 range, and 11 were clearly intermediate. There was a tendency for the five plumage areas used to obtain the index to be positively correlated within individuals. This suggests that the genes regulating plumage form and color act generally rather than upon restricted areas.

Fourteen of the males were placed on a pond with other waterfowl, observed during courtship, and often photographed. Five components of the mating displays which most clearly differentiate mallards and pintails were recorded and scored for 11 individuals for whom adequate film footage was available. A behavior-scoring system was devised in which a 0 denoted a typical mallard and a score of 15 indicated a typical pintail. The average for the group was 9.1, which might be interpreted as overall dominance for pintail behaviors, but the range of 3–15 indicated wide individual differences.

It is interesting to look at individual components. Males of both species during courtship point their bills at a specific female and utter a call. In mallards the pointing is very brief; pintail males hold at attention noticeably longer. Three of the hybrids showed typical mallard bill-pointing; it was completely absent in the other eight. Nod-swimming, another typical mallard pattern, but absent in pintails, involves vigorous swimming in semicircles with head and neck held low. Hybrids showed all degrees of expression of this display. The burp, a pintail vocalization not found in the mallards, was apparent in only 2 of the 11 hybrids; it was either present or absent, never intermediate. The down–up, considered a ritualized form of drinking, consists of lowering the bill into the water briefly and then uttering a monosyllablic whistle. It is characteristic of mallards, but not of pintails, and occurred rarely in the hybrids. These behavioral components were inherited independently. Bird U, for example, courted with a mallard *down–up* and *nod-swim,* but executed *bill-pointing* and *facing* like a pintail.

Especially interesting is the positive correlation of .756 between the behavioral and plumage indices ($p < .01$). This relatively high value suggests that some of the genes affecting plumage may also be involved in regulation of behavior. The result is quite different from Scott and Fuller's (1965) finding that the physical phenotype of F_2 hybrids between two dog breeds, basenjis and cocker spaniels, was completely unrelated to their performance on behavioral tests.

SUMMARY AND CONCLUSIONS

The data I have reviewed so briefly are typical of those available to a behavior geneticist evaluating the contribution of his specialty to communicative behavior. On the developmental side it seems clear that communication patterns, like other behaviors, are the outcome of a series of transactions between organism and environment. The experiments on acquisition of bird songs are particularly instructive. We can neither ascribe their exquisite patterning to instinct nor to the kinds of learning whose laws have been derived from rats in mazes and shuttle-boxes or pigeons in Skinner boxes. Nor do we find one model that fits all species; the rules for each must be learned by experiment.

In the main, observations on communicative behavior in hybrids indicate that the components are inherited independently; and that, in general, differences between closely related species are ascribable to multiple factors rather than to segregation at a single locus. These conclusions are based upon very few cases and are thus highly tentative. Their implication is that there is no master gene integrating the components of a complex pattern, courtship for example, into an effective whole. Such integration as we see is the outcome of a long evolutionary history that has resulted in a coadapted gene complex (Dobzhansky, 1962). Hybridization breaks up the complex—smaller blocks of genes that still hang together are capable of organizing separate components—but the absence of coordination is disastrous for mating success.

EVOLUTION IN SENDER AND RECEIVER

The comparative methods of the ethologist have produced much evidence pertinent to the subject of this volume. Most ethologists agree that display behavior is derived from preparatory movements or from rapid oscillations in behavior attributable to conflict. We might consider such generalized behavior as a wide-band transmission to a receiver that is rather narrowly tuned to certain features of its environment. If it is advantageous to the sender that the receiver respond appropriately to the transmission, there will be selective pressure to emphasize the more salient (from the viewpoint of the receiver) aspects of the intention or conflict behavior.

The process may also be looked at in terms of a narrow-band transmitter with a broadly tuned receiver. Now, if the receiver benefits from attending to particular messages and neglecting others as "noise," there will be selective pressure to develop receptors and associated neurons tuned to the sender's output. The two processes are not mutually exclusive, and it seems to me that both have been effective. They might—as appears in crickets—utilize the same genes and central nervous system characteristics.

There would seem to be strong selective pressures toward making communicatory behavior unambiguous and easily recognized. This requirement would be largely independent of local factors; therefore genetic variation within a species should be small, possibly close to zero. Any marked deviation would, in fact, act as an isolating mechanism and possibly lead to new species formation. If true, one must conclude that our usual intraspecific genetics is unlikely to make major contributions to the evolution of communication. A further deduction is that communication variants such as regional dialects of bird songs and parent–offspring recognition calls will usually be found to be nonheritable. Genes have done their part when they program the nervous system to produce the basic species repertory, leaving the individualization process to the milieu. None of these hypotheses is based upon sufficient data to warrant generalization. In particular we lack good genetic studies of the development of communication in mammals, where the patterns appear less stereotyped than in insects, fish, and birds. What an opportunity!

REFERENCES

Beer, C. G. 1970. Individual recognition of voice in the social behavior of birds. *Advances in the Study of Behavior, 3,* 27–74.

Bentley, D., & Hoy, R. R. 1974. The neurobiology of cricket song. *Scientific American, 231*(2), 34–44.

Clark, E., Aronson, L. R., & Gordon, M. 1954. Mating behavior patterns in two sympatric species of Xiphophorin fishes: Their inheritance and significance in sexual isolation. *Bulletin American Museum of Natural History, 103*(2), 135–226.

Dobzhansky, T. 1962. *Mankind evolving.* New Haven, Connecticut: Yale Univ. Press.

Gottlieb, G. 1971. *Development of species identification in birds.* Chicago: Univ. of Chicago Press.

Hirsch, J. 1967. *Behavior-genetic analysis.* New York: McGraw-Hill.

Konishi, M. 1963. The role of auditory feedback in the vocal behaviour of the domestic fowl. *Zeitschrift für Tierpsychologie, 20,* 349–367.

Lade, B. I., & Thorpe, W. H. 1964. Dove songs as innately coded patterns of specific behavior. *Nature, 202,* 366–368.

Locke, J. 1690. *Essay on the human understanding.* Reprinted. New York: Dover Publications, 1959.

Lorenz, K. Z. 1958. The evolution of behavior. *Scientific American, 199*(6), 67–78.

Marler, P. 1967. Comparative study of song development in Emberizine finches. *Proceedings of the 14th International Ornithological Congress.* Oxford: Blackwell. Pp. 231–244.

Marler, P. 1970. A comparative approach to vocal learning: Song development in white-crowned sparrows. *Journal of Comparative and Physiological Psychology Monograph, 71,* 1–25.

McKay, D. M. 1972. Formal analysis of communicative process. In R. A. Hinde (Ed.), *Non-verbal communication.* Cambridge, England: Cambridge Univ. Press. Pp. 3–25.

Moltz, H. 1965. Contemporary instinct theory and the fixed action pattern. *Psychological Review, 72,* 27–47.

Nottebohm, F. 1970. The ontogeny of bird song. *Science, 167,* 950–956.

Scott, J. P., & Fuller, J. L. 1965. *Genetics and the social behavior of the dog.* Chicago: Univ. of Chicago Press.

Sharpe, R., & Johnsgard, P. 1967. Inheritance of behavioral characters in F_2 mallard and pintail ducks. *Behaviour, 27,* 259–272.

Thorpe, W. H. 1958. The learning of song patterns by birds, with especial reference to the song of the chaffinch, *Fringilla coelebs. Ibis, 100,* 535–570.

Thorpe, W. H. 1972. Vocal communication in birds. In R. A. Hinde (Ed.), *Non-verbal communication.* Cambridge, England: Cambridge Univ. Press. Pp. 153–176.

Tinbergen, N. 1960. The evolution of behavior in gulls. *Scientific American, 203*(6), 118–130.

von Hörmann-Heck, S. 1957. Untersuchungen über der Erbgang einiger Verhaltensweisen bei Grillen bastarden. *Zeitschrift für Tierpsychologie, 14,* 137–183.

Chapter

3

GENETIC VARIATION AND THE EVOLUTION OF COMMUNICATION

J. P. Scott
Bowling Green State University

The most significant advances in evolutionary theory and fact during the present century have come through the science of genetics which originated 75 years ago with the rediscovery of Mendel's paper. As the mechanisms of transmission of the units of heredity were discovered, it became possible to develop genetic theories of evolution that were superior to those that had been based only on comparisons between external and internal appearances or phenotypes. Geneticists, however, have for the most part paid little attention to the evolution of behavior, whereas the ethologists, who have made the largest contribution to evolutionary thought in this area, have been largely unaware of genetic theories, basing their ideas upon comparisons between phenotypes. In this paper I shall make a first attempt to integrate these two fields.

THE GENETIC THEORY OF EVOLUTION

The Concept of Systems

This is a basic biological concept that provides a theoretical basis for the unity of all biological phenomena. The simplest definition of a system is that it is a group of interacting entities. This general definition will apply to entities that are as unlike as planets and genes. Living systems can be differentiated from nonliving ones in many ways, but one of the most important is that, unlike nonliving ones, living systems tend to become increasingly organized, i.e., they are negentropic.

In these terms, evolution may be defined as a change in system organization that persists for more than one generation—thus differentiating it from development, which is a change in organization that persists for only one generation. There are three kinds of biological systems in which long-time changes in organization can occur. These are ecosystems, social systems, and, of course, genetic systems. It is the last sort of change with which genetic theories of evolution are primarily concerned. From the genetic viewpoint, evolution is change in the organization of genetic systems.

There are two kinds of genetic systems. The first is the genotype of an individual organism. The interacting entities are the genes, and they form an enormously stable system which is almost impossible to disrupt or modify without destroying it. In addition, all of the genes found in a given species may be considered as a larger system which we call the gene pool. This is, in one way, a purely theoretical concept, since the actions of genes are always expressed within genotypes. It has reality, however, in the sense that the genotypes can be considered as more-or-less random samples drawn from a common pool. Like the genotype, and indeed like all living systems, the gene pool resists changes in organization and has stability.

Its stability is not absolute, which makes evolutionary change possible.

According to the Hardy–Weinberg principle, the proportions of variable genes in a population remain constant, provided that the population is infinitely large and that its members interbreed at random. This principle simply states that the chromosomal mechanism of genetic transmissions tends to maintain variant genes in constant proportions and may be called the principle of constancy of gene frequency. A genetic theory of evolution must therefore explain how changes in gene frequency take place. Following Wright (1969), we can state that there are three general classes of processes that interact with one another to produce changes in the gene pool of a species.

Mutation

The most important sorts of mutations are changes in the biochemical compositions of genes. Such changes are random in nature with respect to their causes. Under most conditions they occur quite rarely, on the order of one per million per gene locus in any one generation. Their principal effect, then, is to produce random genetic variation. Even though their frequency is low, gene mutations do occur and can bring about genetic change, however slowly.

It is obvious that the smaller the population, the faster the change. In an inbred strain of mice in which the effective breeding population in each generation is 2, only 4 genes at a given locus have to be changed to alter permanently the population; whereas in a population of 1000, there would be 2000 such genes, each of which would have to be changed before the population was completely altered. That such changes actually do occur was demonstrated by Denenberg (1965), who tested two sublines of inbred mice which had been separated for approximately 30 generations. Definite behavioral differences had appeared.

In large populations, the most important evolutionary effect of mutation is to provide a source of variation, i.e., to make change possible rather than to bring it about. Once variation is present, there are two other classes of processes that increase or decrease the speed of change.

Inbreeding

The genetic effect of inbreeding is to produce homozygosity or identity within gene pairs, thus producing a change toward a more pure breeding population in which variation is decreased. The effects are appreciable only in small populations, of which the inbred mouse strain is an extreme example. Thus, within a small local population, the effect of inbreeding is to work against mutation by reducing variation. However, if a species is divided into a large number of small local populations, the effect of inbreeding will be to reduce variation within populations, but to increase variation between them. The effects of inbreeding that take place under these conditions that do not involve selection are known as random drifts in gene frequencies. In short, if a species is composed of small local populations that rarely interbreed, they should drift apart from one another.

Selection

This is a third class of processes and one that is most complex and least well understood. To begin with, it is not a single process, but includes any process that involves differential reproduction between individuals, whether it is brought about by death, by differences in fertility, or in a large variety of other ways. The factors that bring about differential reproduction are known as selection pressures. Each pressure can vary from weak to strong. It may occur sporadically in one generation, or constantly over many generations, depending on the situation. Within a small local population living under constant conditions, its effect should be to reduce variation, but if local populations are affected by different degrees and varieties of selection pressures, selection will increase variation between populations.

The results of selection might thus be indistinguishable from those of inbreeding, except that selection pressures do not act at random. Those individuals whose offspring survive selection are said to be better adapted or more fit than those whose offspring do not survive.

The action of selective pressures in any natural situation is inevitably complex. Such pressures can originate from living systems at any level of organization in addition to those arising from physical conditions. Furthermore, they can affect the survival of systems at any level (Scott, 1968a).

All three of these processes—mutation, inbreeding, and selection—interact with one another to facilitate or inhibit evolutionary change. The details of this interaction are not our concern in the present paper, but it is obvious that the action of all three factors is affected by population size. On general theoretical grounds, Wright has calculated that the most favorable conditions for rapid evolution are those in which a species is divided into relatively small local populations in which there is a moderate degree of inbreeding, modulated by the possibility of a small amount of crossbreeding from adjacent populations, combined with moderate degrees of selection pressures.

The operation of these three classes of factors that bring about evolutionary change is based upon two fundamental phenomena: variation and adaptation. It is the function of this paper to relate these phenomena to the evolution of behavior and to the special case of the evolution of communication.

ADAPTATION AND VARIATION IN BEHAVIOR

An early outcome of the development of genetic theories of evolution was the abandonment of typological thinking. The older naturalists had attempted to describe species in terms of type specimens, using a single individual to describe an entire population. How badly this works can be verified by anyone who attempts to use some of the older keys for the identification of specimens. No two individuals are ever exactly alike and none of them are identical with the original.

The conclusion from genetic theory is that any description of evolving characteristics must be done in terms of populations rather than individuals. Traditionally, populations have been described using means and variances. Some characteristics will vary a great deal, some less, but all show variation to some degree. When we apply this rule to behavior, the case is complicated by the fact that variation is an intrinsic part of the process of behavioral adaptation.

Passive and Active Adaptation

In much of the older evolutionary literature, adaptation is defined as a change in form and structure taking place over many generations and having a useful function. This use of the concept arose because much evolutionary evidence comes from fossils, where only structures remain. The same sort of structural change has been recorded in historical times in the case of changes in color in certain species of English moths. During the course of industrialization in the English midlands, quantities of soot were deposited on the local vegetation, producing the so-called "Black Country." Moths living in this area showed protective coloration similar to the bark of trees, and as these trees became darker so did the moths (Ford, 1964). This kind of adaptive change may be called *passive adaptation,* meaning that the characteristic is acted upon and changed by outside forces.

On the other hand, behavioral adaptation involves changes in behavior and associated physiology as an immediate response to changes in the environment. In fact, variation in behavior is an essential part of the adaptive process. This may be termed *active adaptation.* From an evolutionary viewpoint, behavioral variability itself is adaptive and should be selected for, and so change from generation to generation.

Active Adaptation and the Analysis of Variance

Variation in anatomical characteristics such as are involved in passive adaptation can be analyzed relatively simply into components that are attributable to genetic and environmental factors, thus $V_P^2 = V_G^2 + V_E^2$, i.e., the variance of the population is composed of that variance due to genetic factors plus that due to environmental factors. In anatomical characters it can be assumed that repeated measures on the same individual will give very similar results, because any changes that have been produced by the environment have been made early in development. On the contrary, variation in cases of active adaptation is affected by immediate changes in the environment. The observed variation is composed of components arising from immediate changes in the individual's behavior as well as environmental factors that differentiate his behavior from that of others. Thus, in the case of active adaptation, $V_E^2 = V_W^2 + V_B^2$, separating variance within and between individuals. The net result should be to increase greatly the variance due to environmental factors compared to genetic ones. Furthermore, the variability of an individual's behavior itself may be an adaptive characteristic and thus be affected by selection and genetic factors.

Thus, active and passive adaptation provide very different problems in evolution. With respect to passive adaptation, well and poorly adapted individuals constantly manifest their differences and hence are subject to strong selection pressures. In the case of behavior, this is not necessarily the case. Whether or not behavior is adaptive depends on whether or not it is elicited.

The only way to determine the capacity of individuals to vary their behavior is to make repeated measurements of behavior under standard conditions. When we did this with dogs, we obtained a variety of results, depending on the kind of behavior elicited, the developmental stage of the animals tested, and the nature of behavioral adaptation appropriate to the test situation. First, we found that very young animals show more variation in their responses than older ones (Scott, 1968b, p. 136). Second, when distress vocalization is elicited by separation, repeated exposure causes a decrease in the response, presumably because of habituation, and thus limits the range of variation (Scott, Stewart, & DeGhett, 1974). Third, in any kind of training situation, whether it be forced training as in leash training, reward training as in running to a goal for food, or problem solving, behavior becomes progressively less variable and more standardized. This brings up the problem of the relationship between learning processes and variation.

Effects of Learning on Behavioral Variation

An essential part of the learning process is variation in behavior. Given a problem, an animal at first tries out various solutions and eventually adopts one that provides the best solution. This is then reduced to a firm habit, at which point behavioral variability is greatly reduced. Any selection pressure is thus likely to favor the survival of an older, experienced individual over a younger, inexperienced one, irrespective of the genetic capacities of the two individuals. Because environmentally elicited variation is so important, it follows that differential selection of behavioral capacities is likely to take place only when gross genetic differences exist. In our experiments with dogs we found that many breed differences in behavior could be explained on the basis of one or two genetic factors, probably because under uncontrolled conditions only major gene effects can be identified (Scott & Fuller, 1965). Furthermore, we see that one of the effects of learning processes is to reduce behavioral variation.

Modes of Behavioral Evolution

As seen in the lower animals, such as protozoa, behavior takes the form of relatively stereotyped reactions that are adaptive in special situations. As conditions vary or as a unique situation is presented, the animal may try out these alternate behavior patterns in succession. In such animals, behavior patterns are relatively invariant, their organization having been determined by growth processes, and the adaptive variation is brought about by the use of alternate patterns. Behavior evolved in this fashion is obviously limited. If a new pattern of behavior were necessary in any new environmental situation, behavior would become enormously complex, each animal requiring hundreds or thousands of discrete patterns.

A second mode of behavior evolution found in the more highly organized animals, and particularly in mammals, is that of a few simple patterns of behavior that are widely modified and combinable with one another. To illustrate with examples, the courtship behavior of turkeys involves a complicated series of actions before actual mating takes place. If the male is interrupted at any point, he goes back to the beginning and starts all over again. In contrast, the courtship behavior of dogs and most other mammals is composed of a few relatively simple patterns that do not necessarily occur in a given sequence and that can be evoked or dropped out without necessarily interfering with the final act of mating. Another example of this sort of evolution is the eating behavior of humans, which is first observed as a simple hand-to-mouth reaction in young babies. Eventually this is combined with grasping and placing objects in the mouth, and after considerable experimentation, the babies discover which things are edible and which are not. This very simple pattern can be enormously modified. All sorts of things can be done by the hands on the way to the mouth, including the use of eating utensils.

A third possibility is some combination of both modes of evolution. In fact, we rarely see only one or the other in a given species. Even in human beings there are some relatively invariant patterns of behavior, such as the social smile, and even in birds there are at least some patterns of behavior that are simple and modifiable. Actually, most of the examples of stereotyped behavior described by ethologists are patterns of social behavior, the patterns associated with eating being much more variable.

THE PROBLEM OF STEREOTYPED BEHAVIOR

One of the major concepts of classical instinct theory is that of stereotyped behavior, the so-called fixed action patterns. Assuming that such invariant patterns of behavior do in fact exist, it is useful to consider the conditions in which such a pattern might be adaptive. An obvious answer is that such a pattern can be usefully passed along from generation to generation only in cases where the environment remains constant over long periods of time in terms of generations.

Thus the environment of parasites of warm-blooded animals is extremely constant, not only for the lifetime of the host, but generation after generation. Correlated with this, the behavior of parasites tends to be reduced to simple and mechanical forms and becomes almost reflex in nature. Certain regions in the ocean also provide a relatively constant environment. But for land-living animals, large environmental changes take place seasonally and sometimes from hour to hour. Furthermore, the biotic environment, the ecosystems of which an animal is a part, may also vary from year to year, with some species being very common one year and very rare the next. Indeed, for animals living under these conditions the most stable part of the environment is often the social environment, which persists with very little change generation after generation. This may explain the fact that a large number of fixed action patterns cited by ethologists are patterns of social behavior.

How fixed is a "fixed action" pattern? In most studies, particularly descriptive ones, no object measures have been made and frequently only very small samples of individuals have been studied. From genetic theory we would expect that even in a behavior pattern that appears to be fixed and is found almost universally in a species, there would be genetic variation.

THE EVOLUTION OF COMMUNICATORY SIGNALS

Adaptive Value of Invariant Behavior

Communication through signals is widespread among social species of animals. In fact, communication is a necessary part of any social system as otherwise there can be no interaction among the members. My son once pointed out to me that in addition to behavior adapted to a constant environment, there is another important situation where a behavior pattern that is relatively invariant throughout a species has adaptive value: communication. A signal that is sent in invariant form conveys a clear meaning. We experience this daily in human language communication, where there is constant pressure to speak words clearly and distinctly and in a commonly accepted fashion. Where pronunciations depart too widely from accepted usage, they become unintelligible. The invariant forms of words are, of course, produced by training and habit formation, but such invariant behavior can be organized in other ways, such as growth processes. We would expect that the same principle would apply in more primitive forms of communication, such as signaling behavior—the use of sign stimuli. As an example, many of the social signals used by birds have such apparently invariant properties, presumably because these convey clear meanings.

Signaling Behavior in the Dog

It can be argued that the dog is a poor species in which to study evolutionary processes because it lives in such an artificial environment compared to that of wild animals. Nevertheless, dogs do evolve, and since their normal habitat is that of human society their evolutionary tendencies have considerable relevance to that of the human species.

Dogs use a large variety of visual and vocal signals in communication with one another. For example, the tail held stiffly erect indicates social dominance, while the tail between the legs indicates subordination. Rapid horizontal tail wagging indicates a "friendly" approach. However, the best quantitative data are in the area of vocal signals.

There are relatively few classes of vocal signals in dogs. Barking and growling are associated with the agonistic behavioral system and occur frequently. Howling occurs more rarely and is usually initiated by a solitary animal. Then there are yelps of pain, whines, and the distress vocalizations of the young.

We studied barking in young puppies during 10-min sessions in which two puppies compete for a bone (Scott, 1964; Scott & Fuller, 1965). During this time, most basenji (the so-called African barkless dog) puppies barked less than 10 times, whereas some cocker spaniels might bark several hundred times in an equivalent period. Crosses between the basenji and cocker spaniel indicated that these tendencies are inherited. As used in this situation, the bark expresses a vocal threat to the other puppy, and may be used by either a dominant or a subordinate animal.

Barks differ not only in quantity but quality. A comparison between barks of cocker spaniels and beagles shows that barks of the cocker spaniel are short, sharp, and high pitched, whereas the beagle bark is long, drawn out, relatively low pitched, and in some individuals has an almost musical tone (see Figure 3.1). While the barks of an individual animal vary somewhat in different situations, they are similar enough that individuals can be recognized by their barks.

There appears to be less variation in the howling behavior of dogs, but since this behavior is difficult to evoke, we have studied it very little. The other kind of vocalization that we have studied extensively in puppies is distress vocalization in relation to separation. We have never observed a puppy that would not produce distress calls when separated from familiar individuals and places. The rate of vocalization is easily counted, and the average rate for a puppy placed alone in a strange room is approximately 100 per minute, or 1000 for a typical 10-min test period. Breed differences may be large. Shetland sheepdogs vocalize at a much higher rate than African basenjis; Telomians are somewhat higher; and sheltie × Telomian hybrids higher still (Scott & DeGhett, 1972). There are considerable individual differences among individual puppies of the same breed tested under similar conditions, and these are constant enough so that we can assume that there are individual genetic differences that account for noisy and less noisy puppies.

However, rate alone does not describe the true variation that occurs. A puppy will respond with a mixture of barks, yelps, whines, and howls—in no particular order—will pause for brief periods and start up again, and sometimes vocalize loudly and other times softly. We supposed that this variability had some adaptive significance and at first thought that it might be similar to the vocalizations of certain song birds, where variation in pitch and loudness produce a sound that is difficult to localize accurately, creating a sort of ventriloqual effect. However, blindfolded human subjects had no difficulty in

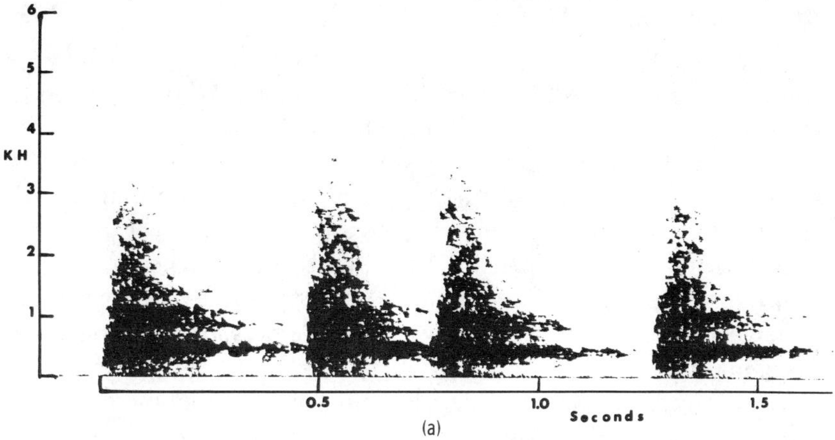

(a)

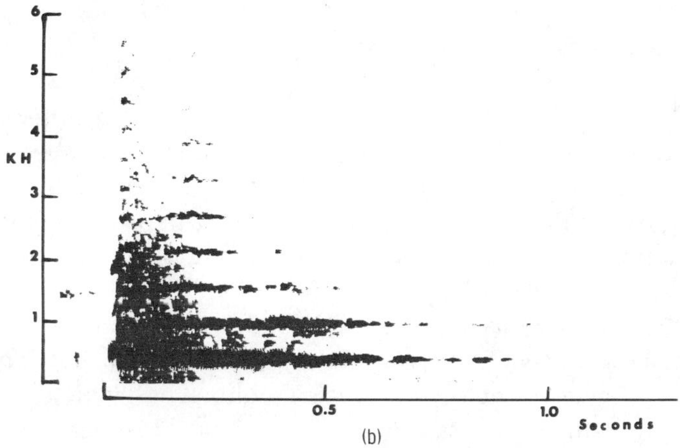

(b)

Figure 3.1 (a) Sonograph of barks by a cocker spaniel. Note that the barks begin with a great deal of noise, followed by a less noisy sound in which a primary tone and its overtones are visible. The upper overtones die out quite rapidly, producing a "Christmas tree" effect. The barks are quite short, not over $\frac{1}{2}$ sec in duration. (b) Sonograph of a beagle bark. This is much longer, almost 1 sec in length. The higher tones persist much longer, producing a more musical effect on the human listener. In addition to these differences, barks given by an individual beagle tend to be more variable than those produced by a cocker spaniel.

pin-pointing the direction from which recordings of puppy sounds were played. It then occurred to us that if the puppy sounds were repeated monotonously and uniformly, rapid accommodation would take place and the hearer would soon pay little attention to them. As it is, the sounds never become monotonous, and I have yet to see an experimenter go to sleep while listening to them. We can advance the theory then that those puppies have survived best that produce variable sounds to which a listening animal or person cannot become accommodated.

This suggests that social signals need not be stereotyped; in fact, that the function of the signal will determine the degree and kind of variation. Considering barks again, the principal function of barking in dogs is as an alarm signal. Barking is expressed as a series of short, sharp sounds rapidly repeated that are very easily localized even when the animal is moving about. Barking dogs running behind a hedge can be located with almost pin-point accuracy. For signals to function in this way they must be short and repeated in close enough succession so that their direction can be easily remembered. The sounds produced in echolocations by other mammals have similar properties. Furthermore, the individual characteristics of the bark make it possible not only to tell where the dog is, but to identify the individual concerned.

This suggests that if signals are used to proclaim identity they should show little variation within the same individual, but considerable differences between individuals. In bird songs, which are often cited as examples of stereotyped behavior, there are sufficient differences between individuals that ovenbirds that have adjacent territories can easily distinguish between strange and familiar birds (Weeden & Falls, 1959). In some species, individual differences are supplemented by regional differences that appear to be acquired by exposure to the sound of an adult bird singing at a critical period in development (Thorpe, 1961).

From this viewpoint, differences in barking rates observed in basenjis and cocker spaniels can be explained as resulting from genetic factors that restrict or extend the range of response. It is possible that in the African jungles in which the original basenjis lived, barking was in some way maladaptive—perhaps by attracting the attention of predators such as leopards—whereas the capacity for long, continued barking of cocker spaniels can be tolerated, if not valued, in an urban civilization.

We may therefore conclude that whether or not a given signal is variable will depend upon the function of the signal. The variability can be relatively large between individuals as it is in the barking of dogs, or relatively large within the same individual as it is with distress vocalization in the same species. I conclude with a plea that those who study signaling behavior should quantify their data and thereby determine its variability and variation. Only in this way can the functioning of the signal be understood.

SYSTEMS AND SIGNALS

Returning now to the genetic theory of evolution, we can begin by relating communication to system organization. Social systems involve behavioral interactions among the individuals that are the entities that compose the system. In turn, the behavior of each individual is organized into subsystems related to particular functions of behavior such as reproduction, conflict resolution, nutrition, etc. Each subsystem is composed of the behavioral patterns that are the units of behavioral evolution.

Some patterns of social behavior involve actual contact, as in sexual behavior and social fighting, but, in addition, there is much social behavior that has a purely signaling function. Some of this is merely supplementary to behavior that carries out fundamental functions, as in courtship behavior when one individual signifies readiness for sexual behavior. In addition, there is a tendency to reduce certain behavior patterns to signals. This has particular importance in agonistic behavior, where, as a result of actual fighting, the behavior between two individuals becomes organized into a dominance relationship in which fighting is reduced to patterns of threat and avoidance. Such reduction has been termed *ritualization* and is brought about by both evolutionary and developmental changes in organization.

Mutation

From genetic theory we can predict that in any species mutations will occur that affect any behavior. Thus, in a given population we would expect that genetic variation will always be present with respect to signals used in communication. There is always some variation present, and therefore there is always the potential of evolutionary change in communication systems. In addition to variation produced by mutation, there is always present in behavior the process of behavioral adaptation that involves variation. As we have seen, such variation may be greater or lesser in degree, depending on the function of the behavior involved, but it is always present.

Inbreeding

Inbreeding and the resulting genetic change has particular relevance to communication. Social organization among nonhuman animals usually involves small local groups because most animal species cannot keep in touch with one another over long distances or over long periods of time. Such local groups are conducive to inbreeding and thus to changes producing differences between local groups. As applied to communication itself, this should result in the evolution of the equivalent of local dialects. If changes in communicatory behavior proceed far enough, they can become isolating mechanisms and thus eventually result in the formation of new species.

Selection

Selection pressures can affect the survival of systems at any level of organization. Furthermore they may result from the activities of systems at any level. This implies that there can be differential selection between different social systems, and one of the ways in which social systems can differ in adaptive function is in the efficiency of communication. In addition, social selection within a social system should favor the survival of offspring of individuals having more efficient signaling skills. These considerations have great importance for the evolution of human language, which not only facilitates behavioral coordination between individuals, but also the teaching of skills relating to the use of tools, which are so important in human adaptive behavior to the physical and biological environments.

MAJOR CONSEQUENCES OF THE GENETIC THEORY OF EVOLUTION

Considered from an overall viewpoint, genetic theory supports certain major conclusions. One of these is that all species are evolving independently. Once two species have become genetically separate they must proceed along different paths, perhaps at slower or faster rates, but always in somewhat different directions. There is no hope then of discovering living nonhuman species that resemble our human ancestors. Even our closest biological relatives, the chimpanzees, are not evolving into people, but into chimpanzees. Furthermore, recent evidence from biochemical comparisons indicates that chimpanzees are evolving as rapidly as the human species on the level of changes in large protein molecules, whatever may be the case with respect to form and behavior (King & Wilson, 1975).

Human and Nonhuman Communication

The major method of communication in the human species is that of verbal language, which often leads us to forget that we not only have the capacity for other kinds of signaling behavior, but also use it. It is with such nonverbal signals that the most direct comparisons can be made with other species.

One of the most studied nonverbal signals is the human social smile. As I have pointed out elsewhere, this behavior has a close similarity in function to tail wagging in dogs, more precisely, the familiar pattern of rapid horizontal tail movement in these animals. Both patterns appear very early in development, both have a purely signaling function that conveys a friendly and somewhat subordinate attitude, and both persist throughout life.

I know of no other behavioral signals in any species that have similar functions. This suggests an alternative theory for the evolution of the social smile from that proposed by Pitcairn and other primatologists, which states that human smiling has been evolved from the fear grimace seen in some other species of social primates. I would like to propose a developmental theory: The human social smile had its original function—and still has a primary one—in infantile behavior, as a social response that is reinforcing and rewarding to caretakers. In contrast to the grimace hypothesis, the smile of an infant primarily reveals that it is toothless and, hence, not dangerous. At the very least, we must consider the alternative hypothesis (which is not at all impossible by genetic theory) that this behavior is not derived from pre-existing behavior, but is novel. This view is supported by the fact that we also have a human behavior pattern that is not only directly analogous, but similar enough to be homologous to the fear grimace—the human snarl, which involves tooth gritting. Indeed, Hutchinson (1972) and others have used the contraction of the jaw muscles as an index of a human aggressive response.

In addition to these problems of the description and function of human nonverbal signals, there is the additional comparative problem that Fouts has raised; namely, whether other primates that do not use verbal language have the capacity to develop a system of communication through visual signals that would have something like the complexity and organization that we see in human language. If this proves to be the case, and even if such capacity turns out to be present only in a moderate degree, it would indicate that prelingual human beings had potentialities that could be utilized in connection with verbal expression. It also raises many interesting problems regarding the interaction of these two modes in the processes of human communication.

CONCLUSIONS

In this paper I have tried to bring out some of the implications of genetic theory with respect to the special problem of the evolution of communication. The ethological descriptive approach to behavioral evolution has emphasized one of the two fundamental phenomena of evolution, that of adaptation. Ethologists are thus quick to propose hypothetical functions for any newly discovered behavior pattern. This has reintroduced a valuable element into the study of behavior, the emphasis on function. Genetic theory, however, leads to the conclusion that behavior can occur without being functional; i.e., new characteristics can occur in evolution without having any use. Therefore, in examining a behavior pattern one must always consider the null hypothesis—that it has no function.

On the other hand, ethologists have neglected the study of a second basic evolutionary phenomenon, that of variation. Their approach has been essentially typological: to postulate that behavior occurs in an invariant fashion throughout a species. Genetic theory states that this is highly unlikely. Even where there may be strong selective forces acting against variation, accidental mutations do occur. Furthermore, an essential part of behavioral adaptation is variation. Therefore, any adequate study of the evolution of behavior must include a study of variation. Even those relatively superficial studies that have been done so far indicate that in the process of communication the degree of observed variation is correlated with function. Communicatory signals can vary both in form and frequency, and unless we have this information we cannot adequately understand either the function or the evolution of signaling behavior.

REFERENCES

Denenberg, V. H. 1965. Behavioral differences in two closely related lines of mice. *Journal of Genetic Psychology, 106*, 201–205.

Ford, E. B. 1964. *Ecological genetics.* New York: Wiley.

Hutchinson, R. R. 1972. The environmental causes of aggression. In J. K. Cole & D. D. Jensen (Eds.), *Nebraska symposium on motivation.* Lincoln, Nebraska: Univ. of Nebraska Press. Pp. 155–181.

King, M. C., & Wilson, A. C. 1975. Evolution at two levels in humans and chimpanzees. *Science, 188*, 107–116.

Scott, J. P. 1964. Genetics and the development of social behavior in dogs. *American Zoologist, 4*, 161–168.

Scott, J. P. 1968a. Evolution and the domestication of the dog. In T. H. Dobzhansky (Ed.), *Evolutionary biology.* Vol. 2. New York: Appleton-Century-Crofts. Pp. 243–275.

Scott, J. P. 1968b. *Early experience and the organization of behavior.* Belmont, California: Brooks-Cole.

Scott, J. P., & DeGhett, V. J. 1972. Development of affect in dogs and rodents. In T. H. Alloway, L. Krames, & P. Pliner (Eds.), *Communication and affect.* New York: Academic Press. Pp. 129–150.

Scott, J. P., & Fuller, J. L. 1965. *Genetics and the social behavior of the dog.* Chicago: Univ. of Chicago Press.

Scott, J. P., Stewart, J. M., & DeGhett, V. J. 1974. Critical periods in the organization of systems. *Developmental Psychobiology, 7,* 489–513.

Thorpe, W. H. 1961. *Bird Song: The biology of vocal communication and expression in birds.* Cambridge: Cambridge Univ. Press.

Weeden, J. S., & Falls, J. B. 1959. Differential responses of male ovenbirds to recorded songs of neighboring and more distant individuals. *Auk, 76,* 344–351.

Wright, S. 1969. *The theory of gene frequencies.* Chicago: Univ. of Chicago Press.

Chapter

4

EVOLUTION OF COMMUNICATION PATTERNS IN ANIMALS[1]

BENSON E. GINSBURG
The University of Connecticut, Storrs

THE PROBLEM

It has been said of complex structures that it is difficult to conceptualize the possibility that they could have evolved by the slow and gradual accumulation of very slight differences guided by natural selection. These differences, if slight and random, could hardly be expected to confer a selective advantage, or to display, in their initial stages, a prechanneled orthogenetic tendency whose integrated increments over time would result in an organ made up of

[1] Supported by an award from the Grant Foundation, Inc.

complexly interacting elements performing at a highly adaptive level, such as an eye. Behavior, instinct, or habit offers the same conceptual difficulty. Of what use is human speech viewed from the articulative side, if the capacity to decode it is lacking? On the other hand, how could a decoding mechanism ever develop out of some primitive precursor until there was a complex signal to decode? Since the two functions are separable in the brain, the possibility that the articulative function and the decoding function are two aspects of the same mechanism does not seem tenable, either.

Explanations in the Darwinian Mode

The explanation of the biological origins of complex structures and behaviors was a problem from the time that organic evolution became an organizing principle in biology. Darwin, in *The Origin of Species*, wrote:

> To suppose that the eye with all its inimitable contrivances for adjusting the focus to different distances, for admitting different amounts of light, and for the correction of spherical and chromatic aberration, could have been formed by natural selection, seems, I freely confess, absurd in the highest degree [p. 133].

He continued, however,

> that if numerous gradations from a simple and imperfect eye to one complex and perfect can be shown to exist, each grade being useful to its possessor, as is certainly the case; if further, the eye ever varies and the variations be inherited, as is likewise certainly the case; and if such variations should be useful to any animal under changing conditions of life, then the difficulty of believing that a perfect and complex eye could be formed by natural selection, though insuperable by our imagination, should not be considered as subversive of the theory [p. 133].

This idea finds further support from the evidence that inherited variations continue to occur, even when the eye has been perfected in the evolutionary scheme, and that these latter variations, if unopposed by selection, can result in the degeneration of the eye, as in the case of cave-dwelling organisms, where vision serves no function, and degeneration of the eye occurs with time. Darwin applied these same principles to behavior. With respect to both morphology and behavior, he argued that abrupt, discontinuous adaptive changes are not likely to occur; nor, if they did, would they constitute the basis for further evolution. In the first place, the changes that are needed to form a complex entity include many systems, all of which must be coordinated and must develop in relation to one another. Further, a number of individuals exhibiting the same modifications must occur in the same place at the same time, or the macromutation will be lost to the population.

> He who believes that some ancient form was transformed suddenly through an internal force or tendency into, for instance, one furnished with wings, will be almost compelled to assume, in opposition to all analogy, that many individuals varied simultaneously. It cannot be denied that such abrupt and great changes of structure are widely different from those which most species apparently have undergone. He will further be compelled to believe that many structues beautifully adapted to all the other parts of the same creature and to the surrounding conditions, have been suddenly produced; and of such complex and wonderful co-adaptations, he will not be able to assign a shadow of an explanation. He will be forced to admit that these great and sudden transformations have left no trace of their action on the embryo. To admit all this is, as it seems to me, to enter into the realms of miracle, and to leave those of science [Darwin, *The Origin of Species*, p. 183].

In *The Descent of Man,* Darwin considered the problem of the evolution of systems of communication, including complex ideational capacities, and the ability to communicate these. In Chapter III, he states that it is his objective to show "that there is no fundamental difference between man and animals in their mental faculties . . . as man possesses the same senses as the lower animals, his fundamental intuitions must be the same [p. 446]." In that chapter, Darwin considers various components of behavior including instincts, emotions, curiosity, imitation, attention, memory, imagination, and reason. In every one of these domains man shares these capacities to varying degrees with other animals,

> man and the higher animals, especially the primates, have some few instincts in common. All have the same senses, intuitions, and sensations—similar passions, affections, and emotions, even the more complex ones, such as jealousy, suspicion, emulation, gratitude, and magnanimity; they practice deceit and are revengeful; they are sometimes susceptible to ridicule, and even have a sense of humour; they feel wonder and curiosity; they possess the same faculties of imitation, attention, deliberation, choice, memory, imagination, the association of ideas, and reason, though in very different degrees [p. 456].

Examples of reasoning, tool using, and complex communication are set forth. With respect to language, Darwin states

> that which distinguishes man from lower animals is not the understanding of articulate sounds, for, as everyone knows, dogs understand many words and sentences. In this respect they are at the same stage of development as infants, between the ages of ten and twelve months, who understand many words and short sentences, but cannot yet utter a single word. It is not the mere articulation which is our distinguishing character, for parrots and other birds possess this power . . . the lower animals differ from man solely in his almost infinitely larger power of associating together the most diversified sounds and ideas; and this obviously depends on the high development of his mental powers [pp. 461–462]. The fact of the higher apes not using their vocal organs for speech, no doubt depends on their intelligence not having been sufficiently advanced [p. 465].

In further support of the idea of phylogenetic continuity in the behavioral realm, Darwin points out the astonishing similarities in brain structure between man and ape, and between apes and other higher mammals. For him, all of the necessary precursors for man's mental capacities, including reason, imagination, and language are foreshadowed in the ape, in the dog, and in the mammalian organization of the brain, endocrines, and sense organs.

Facts that Will Not Fit

While adopting the same evolutionary set today, many of the details or, better stated, the evidence that Darwin used in support of the idea of organic evolution as applied to behavior are now in dispute. Such was the power of the idea to provide an organizing scaffold on which so many observations could be supported and related that it did not and does not matter that some of the observations were in error. The scheme remains intact. However, in itself, it provides an explanation only at the formal and final levels. The material and efficient causes of communication behavior still remain enigmatic and require elucidation in modern scientific terms.

The Darwinian view that the higher apes do not use their vocal organs for speech because they lack the intelligence to do so is not borne out by the work of Lieberman (1975), who has reasoned, on anatomical and acoustical grounds, that the vocal apparatus of the ape, and, indeed, of Neanderthal man, is not capable of producing human speech as we know it. This probably accounts for the fact that the experiments in which chimpanzees were reared as human children failed to produce even a primitive speaker. At the decoding level, also, such human surrogates were a disappointment. It was not until the attempts at complex vocal communication were abandoned in favor of sign language that another level of communication was reached (see Fouts & Couch, this volume). This, too, appears relatively primitive and supports the notion that, though many fairly complex situations can be comprehended and communicated by the signing chimp, he is still different in kind as well as in degree from any human potential for language.

GENETIC MODELS OF COMMUNICATION BEHAVIOR

What, then, do we know in either a developmental or a phylogenetic sense about the evolution of the capacity for communication? It is surely genetic in that a chimp, regardless of rearing, does not develop the linguistic capacity of a man; but is the genetic substratum analyzable in the sense of identifiable genes making identifiable contributions to behavioral signals and to their decoding within a species? If there are such genetic components, as Darwinian theory would require, is there a separate genetic control for the ability to emit the signal and an independent one for its decoding, or are these two necessary capacities controlled by the same genes or by gene assemblies that somehow adhere in transmission?

The simplest and surest model to effect synchrony is that of unitary control of signal emission and decoding by the same genetic mechanism. On the arthropod side of the phylogenetic tree, where so many genetically built-in automatisms occur, such a unitary mechanism has been found to exist (Bentley & Hoy, 1974). Male crickets call to attract females. In sympatric species it is particularly important that the signal emitted by the male should be appropriately responded to by a female of the same species. In order to accomplish this, the signal must be precise and differentiable, and it must be appropriately decoded. In fact, this is what occurs. In nature the calls emitted by the males and the responses to these by the females are species-specific. Moreover, they remain species-specific even if the crickets are reared in isolation or with the wrong species, and are, therefore, in a formal sense, entirely genetic. It is possible to hybridize species having different calls. Since the hybrids are fertile, they can be backcrossed to each parental species. When such hybrids are produced, the males sing a song which is the same for all of the hybrids, but which differs from the song of the parental species. Each backcross male also has a song unique to that genotype. These hybrid and backcross songs are produced regardless of the method of rearing.

Similarly, the females of each genotype respond selectively to the song produced by the same genotype regardless of the method of rearing. The data indicate that the control of both transmission and reception is polygenic, and that some of the same genes are very likely contributors to both systems. Data obtained from males derived from reciprocal crosses demonstrate that genes on the X-chromosome are involved. Further analyses have been performed to ascertain which aspects of the song are altered genetically and what aspects of the neural mechanism are implicated in changing the pattern. We are not, here, primarily interested in the details of the communication system of the cricket. What has been presented should suffice to demonstrate that variation in signal transmission can be under genetic control; and that this same control extends to the decoding system, and most probably involves some of the same genes, thereby providing a fail-safe mechanism to guard against errors in communication.

For the cricket at least, a number of important principles have been established: We know that the call is a taxonomic character—it is both species- and genotype-specific, as is the capacity to decode it and to make the appropriate response. We know that this "language" is independent of experience. Put another way, there are few degrees of freedom between the genetic encoding for the emitting and receiving mechanisms and the resulting behavior. On the genetic side, the basis for this type of communication is polygenic and there appear to be no dominant genes involved. The genes are not contained on any one chromosome. Some appear on the autosomes and some appear on the X-chromosome.

Although the analyses are not as yet complete, there is evidence that a similar situation is present in some amphibians, where male calls are also specifically responded to by females and where the hybrid calls are different from those of the parent species (Bogert, 1961). The innate basis for many bird songs has also been much researched. Here the degrees of freedom between the genetic substratum and its expression are greater, since, in a number of instances, it has been shown that the development of the normal song depends upon exposure to other birds and that the lack of such exposure, or exposure to the songs of other species at particular times during development can, in some cases, have an influence on the call that is emitted as well as on the response (Marler, 1957; Marler & Tamura, 1964). Here the situation is no less genetic or innate, but there are simply more degrees of freedom between genotype and phenotype. Given a particular genetic basis, it is possible by varying the conditions under which the organism is reared, to ascertain the outcome. The degree of flexibility and the mode of interaction with environmental variables at given periods of development are thus also specified by the genotype.

EVIDENCE FROM SOCIAL MAMMALS

Fascinating as these examples are, they are not on the major evolutionary pathway to our own language development. In this respect, the social mammals which exhibit the need to communicate and integrate group behavior, and which have sense organs and a nervous system that is on the evolutionary highway leading to our own constitute the more pertinent example. Is communication behavior in mammals species-specific, genetically variable, and, to any extent, independent of experience? Is the evidence in this area in any way consistent with that obtained for the cricket, in which the characteristics on the emitting side appear to be controlled by at least some of the genes that are at the basis for the capacity for decoding the emitted message?

A partial approach to this problem has been achieved in studies of communication in dogs and in their wild relatives. The comparison of wild species with their closest domestic counterparts affords an example of the effects of varying and different selection pressures on characteristics that have had survival value for the wild species. The problem faced in the wild is to develop an adaptive phenotype while maintaining genetic variability. From a teleological point of view, if aspects of the phenotype that are important for survival value vary appreciably, then the variants are less adaptive than the modal form that has been developed under selection pressure. Selection will favor the latter over the former. If, on the other hand, a species becomes genetically static, it loses its evolutionary mobility under changing conditions. Most species have solved the dilemma of the necessity for maintaining both genetic variability and phenotypic constancy by developing a variety of buffering mechanisms such as epistatic effects, dominance, and interactions that modify the phenotypic expression of genes which, in other genetic contexts, can contribute much more to phenotypic variation. Under domestication, where various forms of inbreeding and a relaxation of selection pressure are practiced, or where selection is for other aspects of the phenotype than those favored in the wild, the genetic variance appears to increase as the result of the breakdown of such buffering mechanisms. Under domestication, man prefers the exotic and preserves it. In the wild, the exotic deviant is maladaptive and nature prefers the norm.

CANIDS AS EVOLUTIONARY MODELS

These principles are dramatically exhibited in the genus *Canis.* Domestic dogs exhibit variation for almost every aspect of behavior and morphology that has come under close scientific scrutiny (Stockard, 1941; Scott & Fuller, 1965). Size, color, pelage, bodily proportions, olfactory acuity, vocalization patterns, and emotionality are all so variable that persons speculating on the evolutionary origins of the dog have often assigned a multiple origin to it. Certainly, if we encountered this much variation in the wild, we would, at the very least, consider that speciation was occurring.

In our own experience, close inbreeding in domestic dogs, even within established breeds, reveals a great deal of phenotypic variation. Crosses among breeds that are mismatches for physical characteristics, such as crosses of long-legged and short-legged breeds, or crosses between short-nosed and long-nosed breeds, when carried to an F_2, indicate that the heredity comes in bits and pieces in that every possible mismatch will occur: large bodies with short legs, small bodies with long legs, pushed-in upper jaws with elongated lower jaws, and vice versa (Stockard, 1941). By contrast, five generations of close inbreeding with wolves and coyotes do not reveal such major variation. Perhaps, for the wolf at any rate, this is because a good deal of inbreeding is practiced in the wild as a result of the pack organization. In hybridization studies, the expression of the genetic variance that is known to be there is also much attenuated, as would be expected if buffering mechanisms have indeed been developed in the wild and selected against under domestication.

If we regard the domestic dog as an indicator of what the genome can do, since the genes reveal themselves more readily where they are less buffered, then it may be possible to ask the question of whether the same genes are involved in behavioral signals and in their decoding for communication purposes. The methodology for asking this question may be illustrated by the approach that has already been alluded to: Where the same genes are involved in determining several aspects of phenotype, then those phenotypic characteristics should remain associated in F_2 and backcross generations where the genetic cards are shuffled. If the genes are associated by linkage or in some other way, then this should also be revealed in such crosses by the fact that those aspects of phenotype that depend upon physically associated genes will most often go together after a cross, but where a relatively large group of progeny are involved, dissociation will be possible.

Before applying these principles, it will be instructive to examine some aspects of communication in wild Canids for evidence of genetic variation and phenotypic constancy.

Differences in Communication Patterns among Canids

In a study recently completed in the red wolf in Texas, the taxonomic status of two presumed subspecies, *Canis rufus rufus* and *Canis rufus gregoryi,* was investigated (Shaw, 1975). Three types of criteria were used: The first involved a combination of morphology and habit. Do the two subspecies occur in overlapping ranges and do they maintain their differentiation there? What is the difference in their appearance and can any such difference be objectified in studies of bodily proportion, dentition, and other measures of morphology. There are no unequivocal answers using this approach. *Canis rufus rufus* is, in general, smaller than *gregoryi,* and larger than *latrans,* which also exists in that area. In its intermediacy, it also overlaps both of these populations, so that there are no clear discontinuities that permit the unequivocal classification of any given specimen.

With respect to more subtle and possibly accurate indicators of speciation, such as blood groups and blood protein differences, the findings are also equivocal. This may indicate that interbreeding occurs with both *gregoryi* and *latrans* or that genetic stigmata from more remote ancestry are still being carried along, even though intercrossing no longer occurs or occurs only rarely.

On the behavioral side, the situation is less ambiguous. Wolves and coyotes, when released from traps, typically run away and show very little aggression toward the person approaching and releasing them. *Canis rufus gregoryi*, on the other hand, is typically aggressive under these same circumstances (Shaw, 1975). More interesting from the point of view of communication, is the fact that *gregoryi* shows the same threat behavior that is seen in the gray wolf and in the domestic dog. It bares its fangs, curls its lips, and growls, thus producing a pattern that is generally referred to as snarling. *Canis rufus rufus*, on the other hand, lowers its head, spreads its jaws widely in a gape, and emits a hissing noise similar to that found in *Canis latrans* (the coyote). In our own crosses of dogs with coyotes, the F_1 hybrids typically snarl and do not gape, and both responses are found in the F_2. Since *Canis rufus rufus* is not a coyote, it cannot represent merely a hybrid population, as these would not be expected to show the gape threat in a consistent manner, unless there had been a number of generations of continued interbreeding with the coyote plus selection pressure for the gape, or unless, regardless of occasional hybridization, the two forms had developed from the same ancestral Canid. On the basis of other evidence, Shaw concludes that the latter was most probably the case and that *Canis rufus rufus* should, on these grounds, probably not be regarded as a mere or recent hybrid.

What is of particular interest here is not so much the taxonomic status of the three wild Canid forms being compared, but the fact that communication behavior is a major differentiator among these. Here we have in a mammal what may be the counterpart of the situation so beautifully researched in the cricket. Canids showing the snarl threat can readily be hybridized with those showing the gape threat. F_2 and backcross generations can be obtained and selectively bred. The proportion of occurrences of gape versus snarl in these generations will give some indication of the mode of inheritance. Rearing gapers and snarlers under a variety of conditions, including isolation, may be expected to reveal the degree to which the genetically encoded emitting mechanism also produces an innate bias toward the understanding or decoding of the same signal as against the opposing signal. These experiments are currently underway in our laboratory.

We have had extensive experience with communication behavior of the coyote, the wolf, and the domestic dog in a variety of situations. Beginning with the wolf, there have been a number of studies, including the now classic work of Murie (1944), involving the observation of packs in the wild. These have been bolstered by studies of packs under semicaptive and captive conditions. The latter permit more detailed observation and also generate hypotheses for testing in the wild.

Darwin, in *The Expression of the Emotions in Man and Animals,* has a number of illustrations of the postural attitudes of wolves and dogs in relation to the messages that these attitudes convey. Schenkel (1947) has expanded upon these and provided a veritable lexicon of wolf expressions. More recently the vocal patterns of the wolf have been analyzed in detail and in appropriate behavioral contexts (Thebergé & Falls, 1967).

In our own studies, we have examined aspects of the communication behavior of the wolf, coyote, and dog and have analyzed the former in three contexts: The first is that of the self-tutored pack. We have now examined three such packs originally constituted from young cubs taken from several different litters that were left to develop in a large population field with no tutors from the wild. Under these circumstances, the pack developed an organization that closely paralleled that found in the wild. The social roles that are individuated appear identical, as do the social signals. All of these are a property of the wolf pack as a developing interacting group. Wolves left to grow up with one another do not need to be taught to have a social structure, to play particular social roles, to threaten, to appease, to court, to patrol, to mate, and to engage in the various behaviors that are characteristic of the pack (Rabb, Woolpy, & Ginsburg, 1967; Woolpy & Ginsburg, 1967). In one sense, this is a gross simplification, since they do influence one another by the way in which they behave. On the other hand, there is a certain stereotypy that can be seen from group to group and that we have characterized as "social genetics" (Ginsburg, 1968). It is not the genetic predestination of an individual wolf to become the pack leader. On the other hand, in every organized pack there is a wolf that performs this function. When observing a pack that is about to mobilize itself for some sort of activity, it can be quite clear, once one becomes familiar with the signals, that a new activity is about to take place, what that activity will be, and which members of the pack are going to be involved in what way.

INFORMATION AND AFFECT: TWO LEVELS OF COMMUNICATION

Communication appears to occur at least at two levels: One of these is information transfer and the other is emotional state or affect. Affect is signaled in a variety of ways and, for anyone who has owned a dog, these signals are familiar. Darwin postulated that there is a narrower repertoire and a greater continuity of affect from lower animals to man, and that, therefore, we understand one another better in this mode. In general, the total aspect of the animal is monitored when such an interpretation is made, but there are also specific signs that indicate the general state of arousal (Ginsburg, 1972). Some of these are behaviorally complex and others appear to be simple, autonomic reflexes. Among the former, vigorous tail wagging indicates a general state of excitement. If the tail is held low, this is usually associated with fear, uncertainty, appeasement, and submission. If the tail is high, this is associated with dominance and potential threat. A low slinking posture, begging and licking movements, the position of the ears and of the lips are all associated with affect as well as with signaling specific meaning in context. On the simpler, reflexive side, piloerection, salivation, muscular tremors, changes in pupillary diameter under constant light, and, less obviously, changes in respiration rate and heart rate are simple autonomic responses that are also indicators of the state of emotionality of the animal. The combination of the affective signal and the signal carrying information with regard to specific intent, such as a threatening snarl, provides the total information that must be reacted to. Baring of the fangs and growling is certainly a threat. On the other hand, it is responded to in the total context of the affect of the animal (Ginsburg, 1972). If the tail is held high, the body posture is erect, and the ears are forward, the animal may be expected to follow through if pushed further. If, on the other hand, the tail is low, the bodily attitude is more plastic, and the ears are laid back, the animal, though threatening, may be expected to back off when challenged; or, if challenged slowly and over a gradually closing distance, its changing intent will be signaled. Wolves appear to avoid actual fights by communicating in this way, during which they monitor one another head to head. We have speculated as to the degree of communicative value that inheres in these autonomic signs. Is the yellow eye, for example, of adaptive significance because the diameter of the pupil can be more easily monitored than in a brown eye? During threat encounters, the reaction of members of the pack are often mobilized in two ways: One is that each side may be joined and the threat display is therefore also a summons to individual allies; the other is that, as a result of the individual interactions or those of two groups, decisions are made to yield ground, food, or other prerogatives, rather than to carry the threat through to an attack.

Communication and Social Roles

During the mating season, social bond formation is heightened and a good deal of communicative behavior is displayed, including courtship behavior on the part of the male, and a soliciting-play type of courtship that may be exhibited on the part of the female. Very often triangles are formed in which a subdominant male pursues and courts a particular female, who, in turn, solicits the attention of another male, who may or may not respond. We do not yet know how these affectional bonds are formed and what factors, other than dominance, are being responded to by the female. We do have evidence that strong social bonds are formed between females and the male who was dominant in the pack at the time that the females were approximately 2 years old (Woolpy, 1968). This response to a dominant male does not, however, necessarily result in an effective mating (one involving a tie and, therefore, the possibility of paternity). What is seen is that a female who has accepted another male after courtship and is receiving his sexual attention will often turn on him and leave him to run with the dominant male if an opportunity presents itself. The dominant male, however, while he regularly disrupts matings by threatening other males—often with no obvious overt signals, but simply by coming into their field of vision and approaching—does not usually court such females or mate with them (Ginsburg, 1968).

Social roles and social behavior are established through interactive experiences and are maintained and coordinated by signals that communicate intent. All of the ingredients for the social organization and communication within a wolf pack are self-contained in the sense that a group of cubs left to develop together have the genetic basis for developing all of the appropriate roles and behaviors, although the degrees of freedom between the genetic substratum and the behavior are obviously much greater than in the case of communication among crickets. Nevertheless, under the appropriate conditions, the behavior appears to be insured.

Developmental Factors in Communication

What happens, however, if a cub is reared in isolation and is then introduced to normally reared wolves? We have attempted this with a female cub taken from her mother at birth and hand raised with a male littermate for 3 weeks. After this, the female was isolated until she was 10 months old. At this time, she was placed in a run next to several normal wolves and separated from them by only a chain link fence. At this stage she exhibited all of the appropriate behavioral components, but had not put them together into appropriate patterns. Nor did she seem to be reacting as though the vocalizations, solicitations, and threats from the wolves in the adjacent pen conveyed any meaning to her. The wolves were then placed physically together under restraint. This permitted more complete interactions, while the risk of injury was minimized. Under these conditions, the isolate came to "understand" the social gestures of the normal wolves and within a period of 4 to 5 days was responding to these appropriately. She had, as a result of immediate, controlled, and repeated exposure to normal wolves, learned to interpret their signals and to put her own previously fragmented behavioral components into the appropriate patterns necessary for effective interaction. It is our hypothesis, based on evidence with other animals, that if the introduction to normal conspecifics does not occur soon after isolation, and in a manner that forces interaction, the isolate can become opaque to the affect and signals of the other animals and can develop what can be best termed "superstitious behaviors" that are inappropriate. Such an isolate is much more difficult to rehabilitate.

Canid–Human Interactions

The third arena in which we have encountered wolf communication is in our attempts to socialize adult animals that have not been previously handled (Ginsburg, 1966, 1972). When such animals are approached, they are extremely fearful. They attempt to avoid the handler and they show obvious autonomic signs, including salivation, extreme pupillary dilation, urination, defecation, and trembling. During this phase, they can only be described as frantic if approached. Clearly, they are opaque to the signals of the handler. Under our behavioral regime, which initially involves simply familiarizing the animal to the presence of the handler, this phase of extreme fear eventually abates and the animal becomes attentive to what the handler is doing. It will approach and sniff the experimenter when the two are separated by the chain link barrier. When this stage has been firmly established, the investigator moves inside the barrier, but stays in a sitting or crouching position in one place. The extreme flight reactions and autonomic stress signs are reinstituted, but abate more quickly. Again, during this phase, the animal appears to be entirely opaque to anything but the mere presence of the handler. When this stage has passed, it becomes attentive to what the handler is doing, which at that time is simply sitting still just inside the barrier. Gradually, in our method, the experimenter will begin to make bodily movements, reach out with a hand, seek eye contact, and establish an interactive mode with the animal, but still without any marked change in his physical position inside the pen. Gradually, by easy stages, the handler begins to move about in the end of the run that is furthest from the animal, and finally approaches the animal. With the appropriate spacing and slow approach, the opacity barrier is now gone and the animal is interacting with the handler. It signals its state of affective arousal and its behavioral intent. If an ambivalent threat is evoked, the handler may still approach. Under these conditions, it is possible that the animal will permit itself to be handled, or it may begin to threaten more seriously, in which case there will be a temporary standoff and the handler can exit from the pen. When, on the basis of the animal's response, the handler decides to interact, there is invariably an aggressive phase in which the wolf may give a warning snap and grab at the handler's clothing or at his person with a warning pressure, but seldom with an actual bite. In time, the animal permits itself to be handled and seeks physical interaction with some avoidance, but no aggression. At this point, the handler usually absents himself for varying periods and when he returns, if he is lucky, he will obtain a greeting (we call this the "vacation effect") and will be permitted to interact vigorously with the animal. After this point, socialization proceeds rapidly to the stage where the animal is virtually a pet. During all of this time, the handler has been guided in his actions by the communications received from the animal. The animal, in turn, is very much affected by the way in which the handler conducts himself. Their joint communication patterns are, therefore, essential in establishing the social bonds.

Additional information may be secured by comparing and contrasting the behavior of the wolf with that of the domestic dog. As we have pointed out, the experienced handler or observer can interpret the behavior of a wolf (Ginsburg, 1966, 1972). If observed in a group of wolves, a solicitation for play, a warning to keep away, a call for aggregation of the group, a serious threat, an ambivalent threat, and a general state of arousal are unmistakably clear. In entering the pen of an individual wolf who is partially socialized, the affective state of the animal can be easily assessed. This provides a necessary background for the interpretation of whatever signals are superimposed on the state signals. These specific signals are then interpretable and can guide the handler in the initiation of whatever interactions he desires. In the case of the domestic dog, however, what might appear in the wolf to be an extremely ambivalent threat might be a real one or the converse might be true. Without knowing the individual dog, it is often difficult to interpret intent. It is as though, on the emitting side, the signals have a less necessary connection with the meanings they once had in the wild. On the receiving side, the same thing appears to be true. The isolate wolf does not understand what it is being told by normal wolves when it first emerges from its isolation. We are not certain at this point whether it is because the decoding must in itself be learned (although there is a predisposition for understanding it, since it happens quickly) or whether the affective state of the animal recently emerged from isolation is such that it is simply opaque to signals that it might be able to decode were the opacity to be pierced. In both the untutored pack and the isolate, effective communication with respect to emitting signals and to reacting appropriately to them comes with experience, but this experience appears to be impinging on a genetically primed substrate.

Dog–Wolf Interactions

In attempts to cross-breed dogs with wolves, the domestic bitch often does not appear to understand the courtship behavior of the male wolf. Although she may be in estrous, she does not interpret the head-to-head encounters as courtship and, instead of responding appropriately to these by standing and averting the tail, or by running playfully with the male, she will generally paw at him and attempt to mount. Perhaps the wolf and the dog have simply drifted too far apart in these respects for us to use this as a valid test of understanding the meaning of a bit of phylogenetically important behavior. In actual fact, therefore, we do not know what the degree of biological association is between the signaling and decoding mechanisms. The fact that the signaling behavior in the domestic dog is variable in its meaning for the dog is evidence that the two are genetically dissociable and that it takes selection pressure to keep them together as a substratum which, when acted upon appropriately during development or by the effective management of the critical socializing experience (as in the case of the isolate wolf), provides the biological basis for an effective communication system.

Some additional light on this problem may be obtained from studies on squirrel monkey calls, in which associations between the gene pool from which the monkeys are derived and the calls that they make and react to have been found (Winter, Ploog, & Latta, 1966; Ploog & Melnechuck, 1971). There is, therefore, the possibility of studying this question in the primate.

APPLICATIONS TO MAN

So far as man himself is concerned, we now have some associations between familial patterns of cross lateralization and difficulties in acquiring language skills. This is not true in all cases of cross lateralization, but the evidence appears good that it is true for cross lateralized left-handed boys (Sadick, 1975). In the case of developmental dyslexia in neurologically normal children of above average intelligence, familial tendencies have been described and genetic hypotheses put forward (Hallgren, 1950; Sadick, 1975).

We thus have three models in which it appears feasible to investigate the relationship between the biological substratum for the emitting aspects of communicative behavior and the decoding aspect of the same behavior. These models are the gape versus the snarl threat, in which Canid hybrids can be produced and studied in a context that should make it possible to determine whether, for this piece of communicative behavior, any relationship exists between the tendency to produce a particular signal and the ability to understand it in opposition to a phylogenetically related one. In the case of the squirrel monkey, there is sufficient genetic variation to approach this problem in the way in which it is already being studied by Ploog and his associates. Finally, in the case of our own species, the lateralization of brain function for the decoding of rapid speech in association with other aspects of brain lateralization may provide one avenue of approaching the problem of the genetic interrelations between these two functions. The more extreme case of familial developmental dyslexia may be one open to a biological experiment that will permit us to explore this same relationship with respect to the interactions between visual and auditory processing. We are, therefore, at a stage where the problem is succeptible of analysis and where we may expect to see clearer results in the near future.

REFERENCES

Bentley, D., & Hoy, R. R. 1974. The neurobiology of cricket song. *Scientific American, 231,* 34–35.

Bogert, C. M. 1961. The influence of sound in the behavior of amphibians and reptiles. In W. E. Lanyon & W. N. Tavolga (Eds.), *Animal sounds and communication.* Washington, D.C.: American Institute of Biological Sciences, Publication Number 7.

Darwin, C. 1859. *The origin of species.* (Reprint ed.) New York: Modern Library Edition, Random House, Inc.

Darwin, C. 1871. *The descent of man.* (Reprint ed.) New York: Modern Library Edition, Random House, Inc.

Darwin, C. 1872. *The expression of the emotions in man and animals.* London. (Reprinted by the Univ. of Chicago Press, 1969.)

Ginsburg, B. E. 1966. Social behavior and social hierarchy in the formation of personality profiles in animals. In J. Zubin (Ed.), *Comparative psychopathology, animal and human.* Washington, D.C.: American Psychopathological Association. Pp. 95–114.

Ginsburg, B. E. 1968. Breeding structure and social behavior of mammals: A servo-mechanism for the avoidance of panmixia. In D. Glass (Ed.), *Genetics, biology and behavior series.* New York: Rockefeller University Press and Russell Sage Foundation. Pp. 117–128.

Ginsburg, B. E. 1972. Anxiety: A behavioural legacy. In *Physiology, emotion and psychosomatic illness.* Amsterdam: Elsevier. Pp. 163–174.

Hallgren, G. 1950. Specific dyslexia: A clinical and genetic study. *Acta Psychiatrica, Neurological Supplementum, 65,* 1–287.

Lieberman, P. 1975. *On the origins of language.* New York: Macmillan.

Marler, P. R. 1957. Specific distinctness in the communication signals of birds. *Behaviour, 11,* 13–39.

Marler, P. R., & Tamura, M. 1964. Culturally transmitted patterns of vocal behavior in sparrows. *Science, 146,* 1483–1486.

Murie, A. 1944. *The wolves of Mt. McKinley.* U.S.D.I. Fauna series no. 5. Washington, D.C.: Government Printing Office.

Ploog, D., & Melnechuck, T. 1971. Are apes capable of language? *Neurosciences Research Program Bulletin, 9,* 600–700.

Rabb, G. B., Woolpy, J. H., & Ginsburg, B. E. 1967. Social relationships in a group of captive wolves. *American Zoologist, 7,* 305–311.

Sadick, T. 1975. Language laterality and handedness in relation to reading acquisition: A developmental study. Doctoral dissertation, Department of Biobehavioral Sciences, University of Connecticut.

Schenkel, R. 1947. Ausdrucks-studien an wolfen. *Behaviour, 1,* 81–129.

Scott, J. P., & Fuller, J. L. 1965. *Genetics and the social behavior of the dog.* Chicago: Univ. of Chicago Press.

Shaw, J. H. 1975. Ecology, behavior, and systemics of the red wolf, *Canis rufus.* Doctoral dissertation, School of Forestry, Yale University.

Stockard, C. R. 1941. *The genetic and endocrine basis for differences in form and behavior.* Philadelphia, Pennsylvania: The Wistar Institute of Anatomy and Biology.

Theberge, J. B., & Falls, J. B. 1967. Howling as a means of communication in timber wolves. *American Zoologist, 7,* 331–338.

Winter, P., Ploog, D. W., & Latta, J. 1966. Vocal repertoire of the squirrel monkey *(Saimiri scuireus):* Its analysis and significance. *Experimental Brain Research, 1,* 359–384.

Woolpy, J. H., & Ginsburg, B. E. 1967. Wolf socialization: A study of temperament in a wild social species. *American Zoologist, 7,* 357–363.

Woolpy, J. H. 1968. The social organization of wolves. *Natural History, 77,* 46–55.

Chapter

5

CONCERNING THE EVOLUTION OF NONVERBAL COMMUNICATION IN MAN

THOMAS K. PITCAIRN
Max-Planck-Institut für Verhaltensphysiologie

IRENÄUS EIBL-EIBESFELDT
Max-Planck-Institut für Verhaltensphysiologie

INTRODUCTION

By any reasonable definition of the word *communication,* both animals and men communicate with one another. The behavior of one animal is broadcast to another, who perceives it and modifies his own behavior accordingly. Cherry (1957) defines communication as the "establishment of a social unit from individuals by the use of language or signs" (thus including animals) and Altmann (1967) as being that behavior which produces a change of behavior in the receiver. It is thus the basic premise of this paper that if animals have produced a communicative system that can be shown to have an evolutionary continuity along parts, at least, of the phylogenetic tree, up to and including primates—our nearest relatives—then it would seem more likely and reasonable than not that there is also a continuity of communication systems into human behavior, particularly that part labeled nonverbal communication.

The further implication of this statement is that this continuity must be brought about by some form of information continuity also—the individuals and species must have some common repertoire or information retrieval system. As this can be demonstrated in the case of many animal species, we would thus also expect it to be true for man. That is not to say that there is no cultural component, and we shall later go into some detail about this relationship, but for the moment let us examine the first proposition.

BEHAVIOR AS AN ADAPTATION

When we speak of adaptations, we mean that those structures designated as adaptive fill a role and thus serve to conserve the individual and/or species. Adaptations presuppose an interaction between the organism and the environment and the result is a reflection of facets of this external truth, if you like. This holds true for all adaptations, whether of body form or behavior.

These adaptations are the static form of what Klopfer (this volume) calls changes in species. The total "lineage" of these changes forms the evolutionary tree to the species, again in Klopfer's terminology. If this is so, then the "metaphor" of evolution serves one function in biological theory, that of delineating the changes that *may in the future* take place, because of the prior adaptation of a species. These changes, or adaptations, themselves act to limit further changes, either in the positive sense (e.g., the change between the bared teeth displays in primates and smiling/laughter in man, discussed later), acting as preadaptations,[1] or in a negative

[1] Preadaptation, of course, is only valid as a concept to the species in which the change (to an adaptation) will occur, which may in itself be an adaptation, or *possibly* neutral.

sense where the structure itself has been reduced or is missing and thus no longer open to further changes. This further implies that changes take place within the limitations of structure placed upon them by the prior structure. The series of pentadactyl limbs from reptiles, birds, and mammals include structures modified for various functions (swimming, flying, fast running, digging, etc.) but are composed of the same ground structure—the bones bear the same relationship one to the other. This must then also be true of behavior. The basic relationships involved in behavioral structure must be similar as the behavior changes from one species to another. Two ways of assessing this structure will be discussed further.

In order for an organism to be adaptive it must be in a position to receive information about all environmental changes that are the subject matter of that adaptation. The organism has two possible ways to gain information: first, experientially and second, genetic adaptations resulting from a fluctuation in mutation and selection. Mutation alone is not enough, in that selection pressure controls the direction of any change. The total phylogenetic experiences of a species are stored in the genome and decoded within a process of self-differentiation.[2]

Organisms can also collect and store information from their individual experiences. In that we are discussing adaptations in behavior, these experiences are gained by learning and stored in the central nervous system, and may be collected by active exploration by the organism, or by active teaching, thus becoming a part of tradition. These two processes are equivalent to Scott's (this volume) "passive adaptation" (change from one generation to another, becoming better adapted to survival, exemplified by industrial melanism in moths) and "active adaptation," where an animal varies his own behavior or physiology. The origin of these adaptations, however, has given rise to the whole of the nature–nurture conflict.

[2] By self-differentiation we mean that the behavior changes only in relation to the maturation of structures that support it—thus a behavior may become related to more specific releasers (in the case of smiling, for example), or the performance itself may become better (grasping at objects by an infant). The time relationships of a behavior may only be due to whether the underlying substrata are matured or not.

Today there are extremely few behaviorists who would describe an organism newly entering the world as a tabula rasa, but rather, as biologists have done for a long time, speak of an interaction between genome and the environment—saying in the same breath, however, that it is impossible to establish what concrete contribution the genetic or environmental parts make. We do not entirely accept this defeatism. Developmental physiologists have been able in many cases to separate the components, and ethologists, with a succession of deprivation experiments, have established a relevant method. They tell us about the origin of specific adaptations. One can answer the often-heard objection that as one cannot withhold all experience from an animal, it is thus in principle not possible to demonstrate the genetic determinants (or inborn components) of behavior by simply showing that this is not necessary. Because adaptations are brought about by the pressure of specific environmental "givens," the organism must therefore gain a very specific range of information, and it suffices then to withhold only that specific information.

When anyone has heard several birds of one species singing the same song, he can then ask himself if this specific song is a product of ontogeny or not. In order to answer this question, it is sufficient to raise the bird in isolation from the laying of the egg onwards. If it then produces the adult song, one knows that the information must have been contained in the genome. In what way this is decoded is another question. If anyone wishes to know if the red belly of a stickleback produces an inborn aggressive reaction, then it suffices to raise it in social isolation and present it with a red object. Rearing the animal in total darkness produces a totally different level of disturbance. We know that this produces a degeneration within the retina.

We call behavior elements that in this sense mature without any dependence or experience and that owe their specific adaptability to genetic developmental processes phylogenetically adaptive, or inborn. When we are concerned with movement patterns, this does not mean that these movements as such are inborn, but rather that because of the underlying nervous system and its specific switches with the target organs in relation to the ancestrally determined developmental pathways, they grow out of a process of self-differentiation.

Thus, animals show various behavior features that are dependent for their potential existence, if not expression, on genetic encoding. This is a sine quo non for evolution to proceed, and this insight is hardly new. However, when man comes to be included under this rubric, then many dissenters are to be found, particularly in the fields of anthropology and psychology. As we look back on a long phylogenetic history, which has determined our present day anatomical, physiological, and biochemical status, it would be simply astounding if it were found *not* to affect our behavior also. As we stated earlier, this belief forms the foundation of this paper. Fortunately, there is also considerable evidence for this position.

PHYLOGENETICALLY ADAPTED BEHAVIOR IN MAN

Many research workers have made inventories of the newborn's behavior and found, immediately following birth, that the infant has a whole series of fully functional behaviors. Recent findings of Ball and Tronick (1971) and of Bower (1965, 1966, 1971) further show that infants of only 14 days can immediately interpret certain environmental information. If, when one fixes the infant upright in a chair and projects a dark image in front of him, this object enlarges symmetrically, the infant reacts by turning away the head, covering the head protectively with the arms, and by evidencing a rising heart rate, as though there were an object on a collision course. If the image is made to enlarge assymmetrically, this does not happen. The infant has, to a certain extent, the expectation that these optical stimuli have tactile consequences, without any prior experience. Other experiments also show this—for example, if an image of an object is projected in front of a 15-day-old infant, he attempts to grasp it, although in a poorly directed fashion. This is unexpectedly early. When the child grasps on nothing, he shows signs of discomfort, whining or crying. If, however, he can grasp or touch an object, he remains calm.

Eibl-Eibesfeldt (1973a,b,c,d) has also approached this problem by examining children who have developed under specific sensory

Figure 5.1 Deaf–blind girl Sabine, upper smile and laugh. [Photo from 16 mm film by I. Eibl-Eibesfeldt, 1973b.]

deprivations. Among others he has followed the development of deaf–blind children. Although they have grown up in never-ending silence and night and cannot therefore have any social experience as a result of seeing or hearing, they show to a large extent the normal facial expressions found in other children.

Figures 5.1, 5.2, 5.3, and 5.4 show some of the facial expressions seen. Figure 5.1 is of a normal laugh with upper smile (a smile in which only the upper row of teeth is seen; Grant, 1969), and Figure 5.2 shows the simple smile followed by the wide smile, in which the mouth is not opened. Crying is seen in Figure 5.3. Note particularly Figure 5.3b, in which the sad frown (outer corners of the eyebrows lowered) can be seen with mouth corners down, the normal components of a sad face, and Figure 5.3f, with the tightly closed eyes, still with sad frown, and lips retracted in crying. Figure 5.4 shows the investigation of an object by touch and smell. Perhaps they are indeed oriented by their deprivations to learn by touch, and have thus obtained the information. However, one can show that this

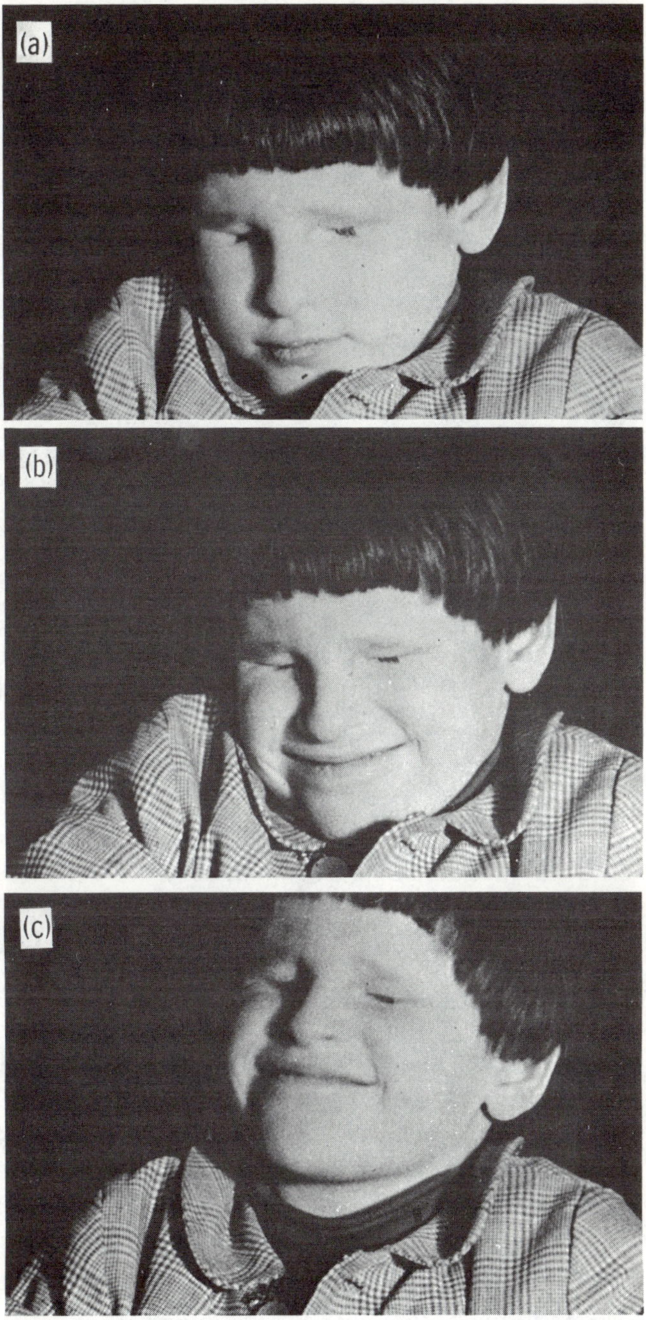

Figure 5.2 (a–c) Sabine, the transition to wide smile. [Photos from 16 mm film by I. Eibl-Eibesfeldt, 1973b.]

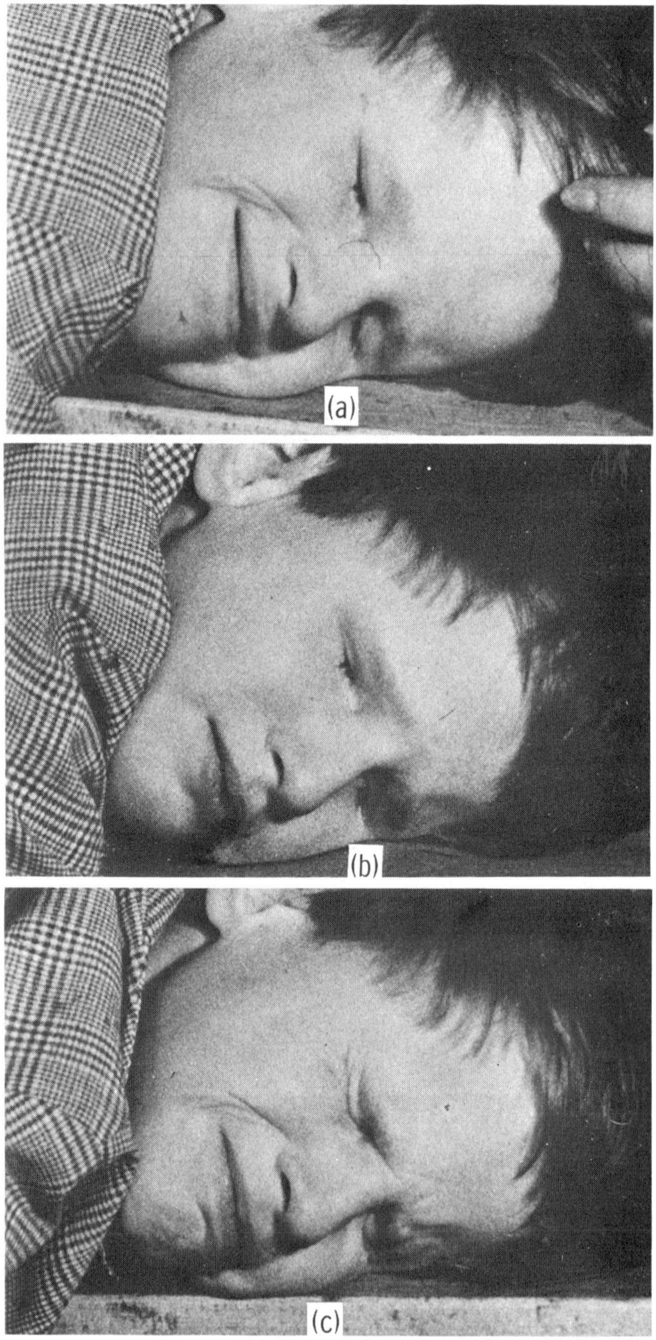

Figure 5.3 (a–f) Sabine, sequence of crying. [Photos from 16 mm film by I. Eibl-Eibesfeldt, 1973b,c.]

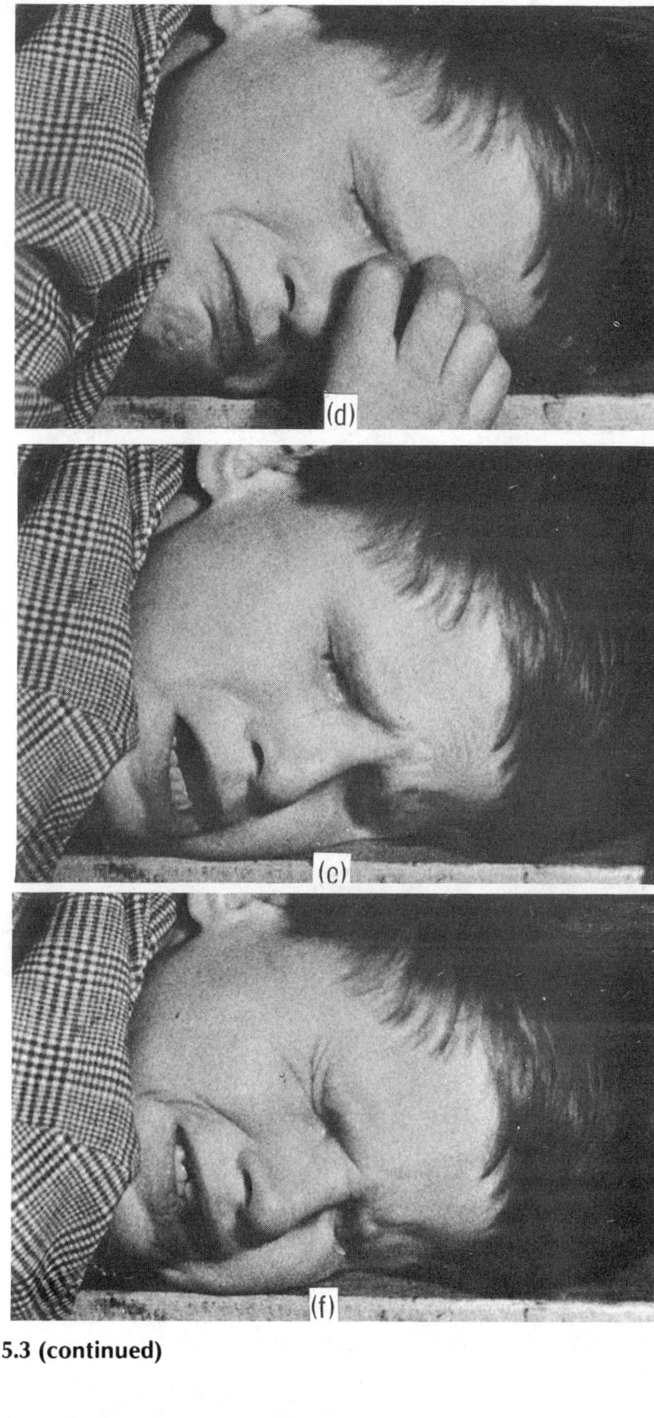

Figure 5.3 (continued)

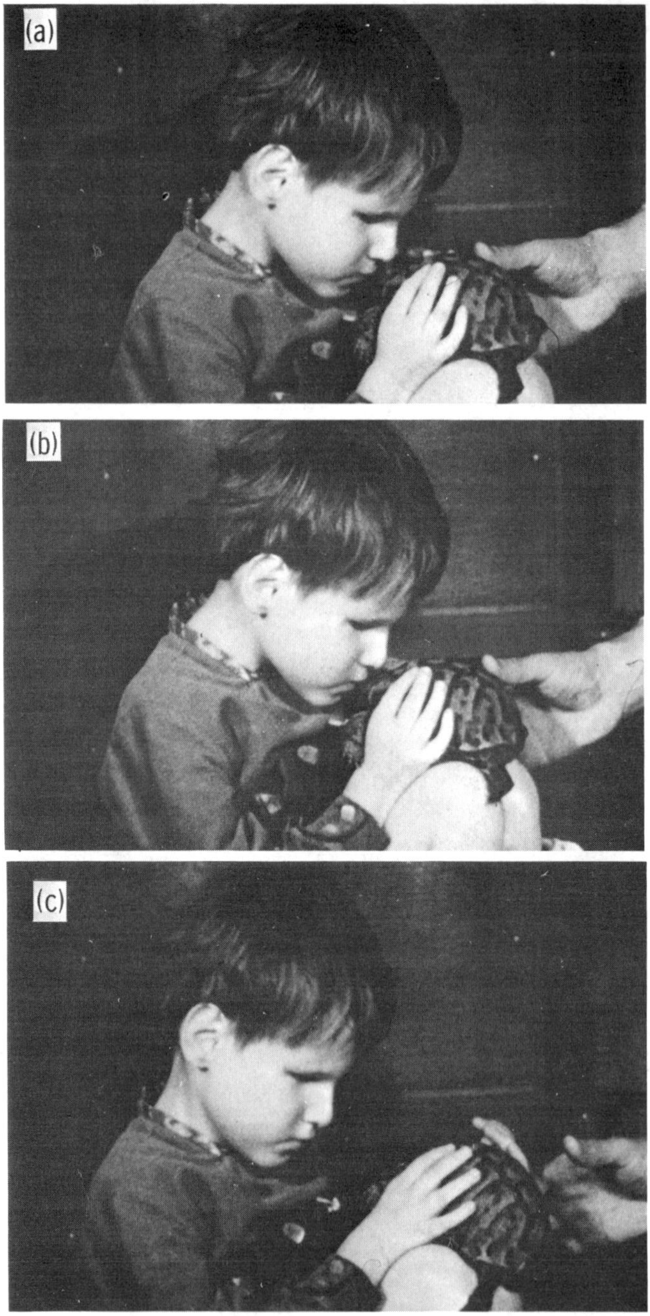

Figure 5.4 (a–c) Sabine, investigation of new object. [Photos from 16 mm film by I. Eibl-Eibesfeldt, 1973b.]

type of learning (which would also require a very specific ability to learn) plays no part by demonstrating the normal patterns of crying, smiling, laughing, and anger even in thalidomide children who are deaf–blind and without arms. The only possibility that remains (outside that of preprogramming) is that the behaviors are shaped through the reinforcing of them when the appropriate form occurs. This also implies, however, that the behavioral elements must in any case be known. We can dispose of this by pointing to highly subnormal children, whom one has great difficulty in teaching to raise a spoon to their mouths, who also show these normal behavior patterns.

Brannigan (1972) has also shown these behaviors in a microcephalic child in whom little or no cortical learning is possible. Thus the idea that human facial expressions are totally learned is not acceptable.

The possibilities of observations with those unfortunates born deaf and blind are, of course, limited. With children born blind only, one can go further, as complex social behaviors can be released by the acoustic channel. When a 10-year-old girl was praised for her piano playing, she blushed. She fixated the observer with her totally blind eyes, then dropped her gaze and turned her head away, and then looked again. The normal nystagmus of the blind was missing when she "looked" at. As this girl cannot see, this behavior must have come from a centrally fixed program. We can see a similarly marked ambivalence in turning toward and away in the coyness behavior of sighted children.

One of us (Eibl-Eibesfeldt) has filmed this behavior in an 11-year-old blind-born boy. When asked if he had a girlfriend, he hid his face behind his hand and "looked" down. This was repeated on three separate occasions, eliciting the same response each time. It seems highly unlikely that this cut-off behavior (Chance, 1962) could be learned.[3]

[3] This is especially so when Chance's hypothesis on cut-off behavior is considered—that these act to cut off incoming behavioral input (signals from the interactor), in this case in the visual mode. This could not here be learned as a control mechanism of arousal, as the signals were never present.

SOCIAL BEHAVIOR AND COMMUNICATION

We have established that certain social behaviors are under so-called genetic control, or are built into the wiring diagram, but is there any difference between social behavior and communication in general? It is certainly very difficult to draw a hard and fast distinction between them, for whatever one individual does exerts an influence on any other interactor. Social behavior is thus the partial and usually reciprocal determination of an individual's actions and interactions. Even when one attempts, as many communication theorists such as McKay (1972) have done, to define communication in such a way as to include some concept of directedness or purposiveness, this does not exclude various general models. Hebb and Thompson (1954) state that the essence of any purposive Communication is that "the sender remains sensitive to the receiver's responsiveness during sending, and by modification of his sending shows that his behavior is in fact guided by the intention of achieving a particular effect in the receiver [p. 739]." However, this would be true of any fight between two fish, or dogs, or monkeys, or men, and thus seems to serve no purpose in clarifying the situation.

Thus, communication patterns of man are composed of elements of social behavior, some of which, at least, can be seen to be genetically controlled and open to prior evolutionary processes. In fact, of course, even culturally determined aspects of behavior are open to evolution, a point we wish to develop later. Let us begin by looking at one particular aspect of human behavior, laughter and smiling. This behavior has been studied by many authors, but van Hooff (1972) has proposed a model of the possible evolution of these behaviors. In earlier work, van Hooff (1962, 1967) had derived a list of facial displays that were common to many primates (with minor variations) such as the bare-teeth open-mouth grimace (Pitcairn, 1975), and the play face (Figure 5.5). These expressions show a certain formal similarity to laughing and smiling, and this is even more marked when one considers chimpanzees. Here, there are four relevant expressions:

a. Horizontal bare teeth. This is very similar to the grimace of other primates.
b. Vertical bare teeth. This involves a vertical retraction of especially the upper lip, with the mouth almost closed, and occurs in relaxed situations.
c. Open mouth, bare teeth. In this expression, full horizontal and vertical retraction of the lips are seen, with the mouth open, occurring also in situations where the animal is calm and relaxed.
d. Play face. This is very similar to the same expression in other primates (Figure 5.5). This is found (also in playing children) associated with a panting "ah-ah-ah" vocalization, and the animals are seen to be in an excited state during vigorous play.

The bare teeth displays were found to be associated with a reduction in aggressive, submissive, and excitement behaviors, with the vertical bare teeth being found especially with a reduction in excitement behaviors and horizontal bare teeth with a reduction in aggressive behaviors. They also differed with the behavior of the partner that preceded them, the former being preceded, of course, by excitement and the latter by aggressive behaviors.

Van Hooff then conducted an enquiry into when a smile or a laugh was thought to be typical by subjects. To cut a long story short, he found that the smile was assessed to be mainly affinitive in form, whereas the laugh was thought to be mainly playful. He could then construct a two-dimensional table, with the two dimensions being increasingly playful and increasingly affinitive, and place the various smiles and laughs seen on it. The playful dimension was seen to include increasing mouth-opening and vocalization, whereas the affinitive (or nonhostile) was composed of increasing lip-retraction, ending in the smile typical of greeting situations, the upper smile, as will be shown later.

What, then, are the connections between this and the bared teeth expressions? A diagram can be constructed to show this (see Figure 5.5).

Both the grin face and the play face occur in virtually the same

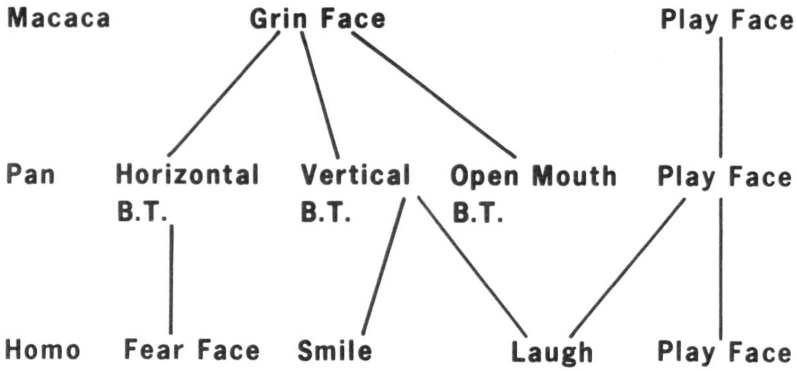

Figure 5.5 The relations of bared teeth faces (abbreviated B.T. in figure) to smiling and laughter. [Modified from J.A.R.A.M. van Hooff, 1972. A comparative approach to the phylogeny of laughter and smiling. In R. A. Hinde (Ed.), *Non-verbal communication.* Cambridge: Cambridge Univ. Press, 1972.]

form in the three groups shown. Laughter, however, as an expression used in play and exhibiting willingness to play (in the broadest sense), is compounded from the play face, with its panting vocalization and the vertical and open-mouth bare teeth displays. In children you can often see the play face intergrading with the broad, open-mouthed laugh. The smile, however, is more affinitive in nature and is derived more from the bared teeth expressions, denoting "I am a friend." This is not to say that the expressions of laughter and smiling are derived directly from macaques and chimpanzees—we did not evolve from a chimp—but rather that as these primates represent our nearest ancestors along the phyletic scale, we can presume that the modifications shown here would parallel those of our direct ancestors.

There is a considerable body of other evidence to show the inborn (and thus open to such phyletic descent) nature of laughter and smiling. Eibl-Eibesfeldt's work, mentioned earlier, on severely deprived children, who nevertheless still showed these expressions, can be related to the early work of Spitz and Wolf (1946), who showed that in 3- to 6-month-old infants a smile could be released by scarecrow faces and distorted grimaces as well as by a normal human face, and of Ahrens (1953), who followed the development in greater detail. Eibl-Eibesfeldt also showed that the response to mimic expressions of the forehead did not appear until 14 months, and that vertical threat wrinkles frightened the children although they had hardly, if ever, seen this expression. Ambrose (1961, 1963) showed that smiling occurred spontaneously also during feeding, changing, and winding. Robson (1967) added visual contact to the effect of smiling as a care-giving releaser (Bowlby, 1958). Eibl-Eibesfeldt (1972) has also shown that smiling can be seen in many different cultures; that is, it has a cross-cultural universality. Ekman (1972), by using a series of still photographs as test objects, demonstrated that the expression of smiling is recognized across many different cultures as being related to the state of mind, or emotion, "happy."

STRUCTURES AND PATTERNS IN BEHAVIOR

The work cited in the preceding paragraph relates to the structural form, and motivation, of particular single behavior elements. However, it is also possible to look at the structure of behavior in total, meaning in this case the relationships seen between all elements of behavior, or how they fit into one another. One way to do this is to look at the probability relations of sequences of behavior—what elements of behavior follow or precede one another most often? One of us (T. K. Pitcairn) has done this for both a macaque species *(Macaca fascicularis)* and for man in an interview situation, and the results make an interesting comparison. Figures 5.6 and 5.7 show the relations between the various groups of behavior [for details of the method see Grant & Mackintosh (1963) and Mackintosh (1966)], in monkey and man respectively. The first point to note is the central position of looking behaviors, divided here into look at another individual and the cut-off gaze behaviors (look away, down, or up). This is true for both species. A second point is the relationship, particularly evident in the macaques (Figure 5.6), between look-at and agonistic behavior (flight, fight, and submission). In the human case the picture is slightly different, as flight behaviors are linked mainly to the cut-off elements. However, as the cut-off elements themselves are somewhat flight-oriented, and the behavior was recorded in an adult interview situation where the level of such behavior would be expected to be low, the difference may not be significant. Certainly among playing children flight is strongly associated with look-at. Pitcairn (1974a) has shown this in pongids.

The third point of interest is the correlation between look-at and social positive behavior, contact and smiles in the case of man. These features show that the structure of the behavior, the way it is organized, is similar in man and monkey. That, of course, is not to impute directly any direct genetic control of such organization—of this we have, as yet, no information. However, it does seem likely that this structure is epigenetic in nature at least, dependent (insofar as those features just discussed are concerned) upon the nature of gaze direction patterns. Looking-at is strongly correlated with threat in most primates—stare, with no other facial expression, is a recognized threat (Angst, 1974; Bertrand, 1969; van Hooff, 1962)—and in most situations to look at another for extended periods is thought

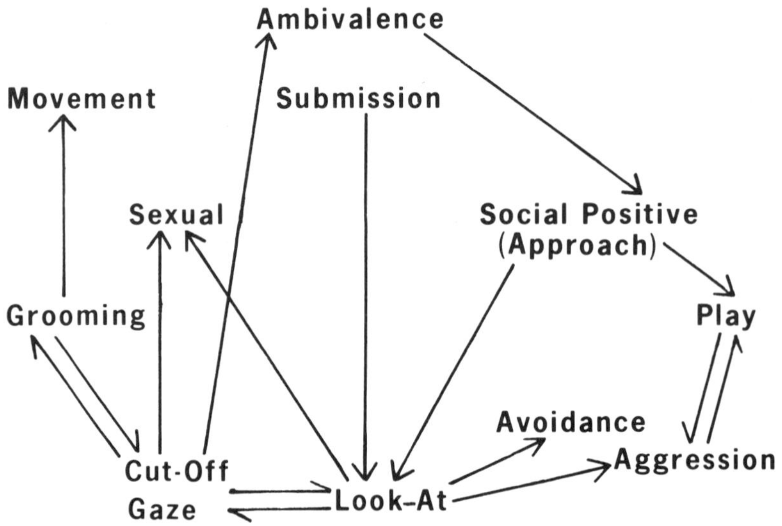

Figure 5.6 The relations of various groups of behaviors in *Macaca fascicularis.* Arrows indicate direction of expected transition. [Modified from Pitcairn, 1975.]

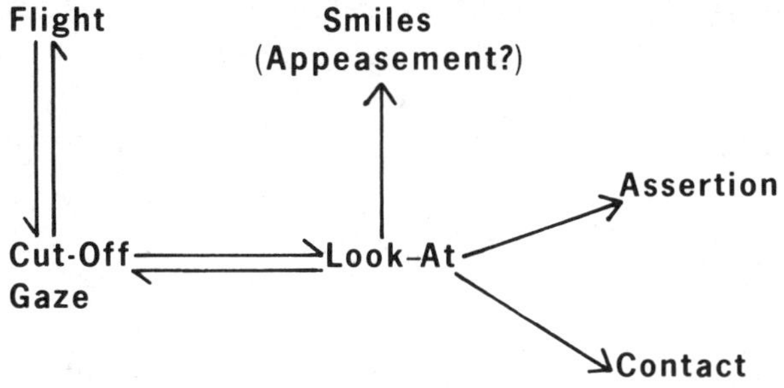

Figure 5.7 The relations of various groups of behaviors in man to one another. Arrows indicate expected direction of transition. [Modified after Pitcairn & Grant, 1972.]

to be at least impolite and may in itself lead to aggression ("What do you think you're looking at!"). However, this behavior *is* modified somewhat in man by the addition of speech. Looking-at is closely correlated with the expectation that the other will continue speaking, whereas one looks away during an interruption or at the beginning of a sentence (Kendon, 1967). Man also is the only animal in whom we find extended face-to-face interactions—in fact, the very physical nature of small groups would seem to us to be built to allow maximal interchange throughout the group. Such groups are built in a circular form, where the "action space," so to speak, of all interactors overlaps. This allows the members to switch most easily from one individual to another. In most primates, contact is mainly limited to side-by-side or ventral to dorsal surfaces, not face-to-face. This has other implications, of course, in terms of the peculiarly fine development of facial expression in man (Kendon, personal communication).

Another behavioral phase in which gaze direction patterns play an important part is that of greeting. By this we mean the particular series of postures and gestures that occur when two individuals meet each other. In humans, greetings, of course, are in some senses extremely variable, depending on such factors as the relative social status of the two individuals, the place of greeting, the time since they last met, and so on. However, in the particular circumstance of one individual approaching another there is one part of the behavior that seems so far to be almost invariant—the eye contact patterns of the approacher. The accompanying photographs, taken from a 16 mm film shot in South Tirol (Italy), show two examples of the same thing (Figures 5.8 and 5.9). They show an individual approaching a waiting "greetee" and giving first the distance salutation—a recognition of the presence of the other—approach, and then close greeting (both terms after Kendon & Ferber, 1973). In the distance salutation of both the man to a friend and the woman to her sister, one can see the head bob (head tossed backward) and upper smile (smile with only the upper row of teeth showing), with direct look at the greetee.

During the approach phase the gaze is averted, to the extent in the sisters' case of looking completely away to the neighboring building (the church). Gaze contact is then once again made in the last approach phase and the close greeting given, followed by a complete cut-off, turning away, before re-establishing contact and continuing the interaction. This pattern has been seen by Kendon and Ferber (1973) as well as by the senior author in various different groups. It can be considered as a bout, or stream, of activity (Pitcairn, 1974b), which, once started, must continue to the end. In other words, once the distance salutation is given, an interactor is committed to proceed to the close greeting and some form of interaction.

It is interesting to compare this pattern with that seen in young infants (Figure 5.10) and primates (Figure 5.11). In the case of an infant about to be picked up, he holds out his arms with direct gaze contact as an invitation, then averts his gaze as he is picked up, then looks again. Thus the pattern is the same in an analogous situation. In young children in kindergartens one can see exactly the same pattern on approach, and also in primates. What does this then mean? This pattern of behavior may be explained on various functional grounds, such that the direct gaze contact would be interpreted by the interactor as threatening, as explained earlier. However, what is important here is, first, that these behaviors are maintained in the population of all primates (including man) examined and, second, that they occur in very young and visually deprived subjects. This latter argues for some central control of the pattern, as it is difficult to envisage how such a behavior could then be learned.

Figure 5.8 (a–f) Greeting of woman to her sister, South Tirol, Italy. Movement toward with close greeting (c), cut-off turn (e), and beginning of interaction (f) (continued on next page). [Photos from 16 mm film by T. K. Pitcairn.]

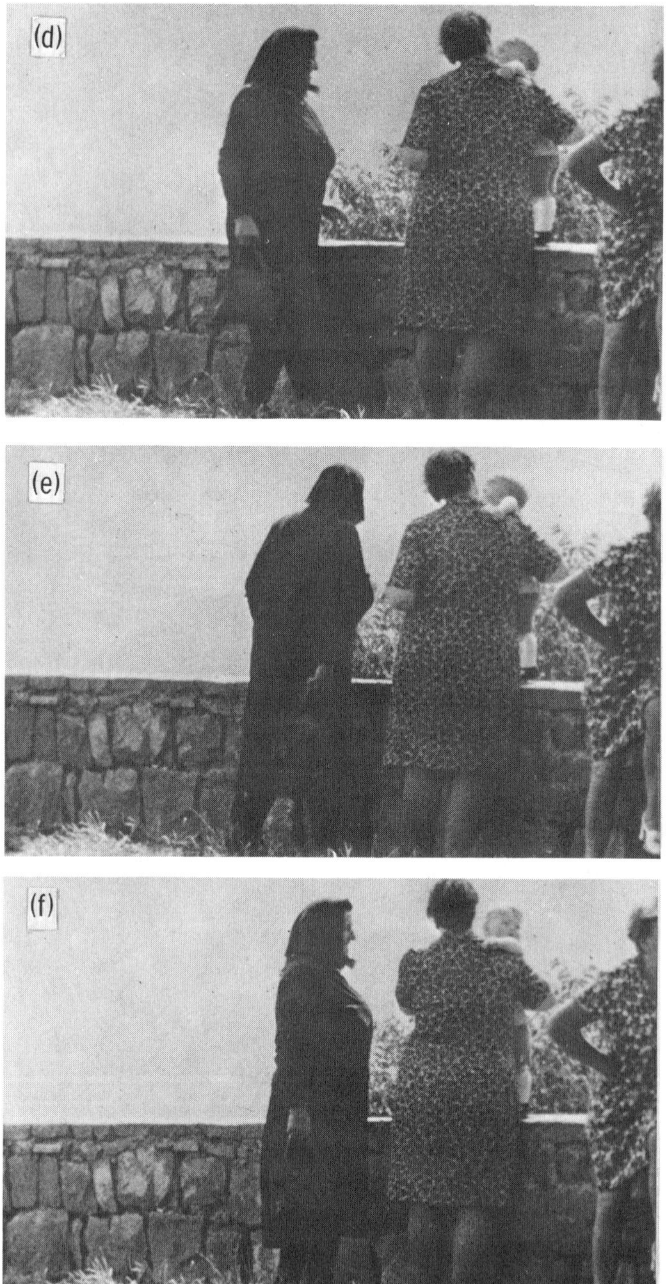

Figure 5.8 (continued)

Figure 5.9 (a–h) Greeting of two men with distance salutation of upper smile (a), and close greeting handshake (b–d), cut-off (f,g), and interaction (h) (continued on next two pages). [Photos from 16 mm film by T. K. Pitcairn.]

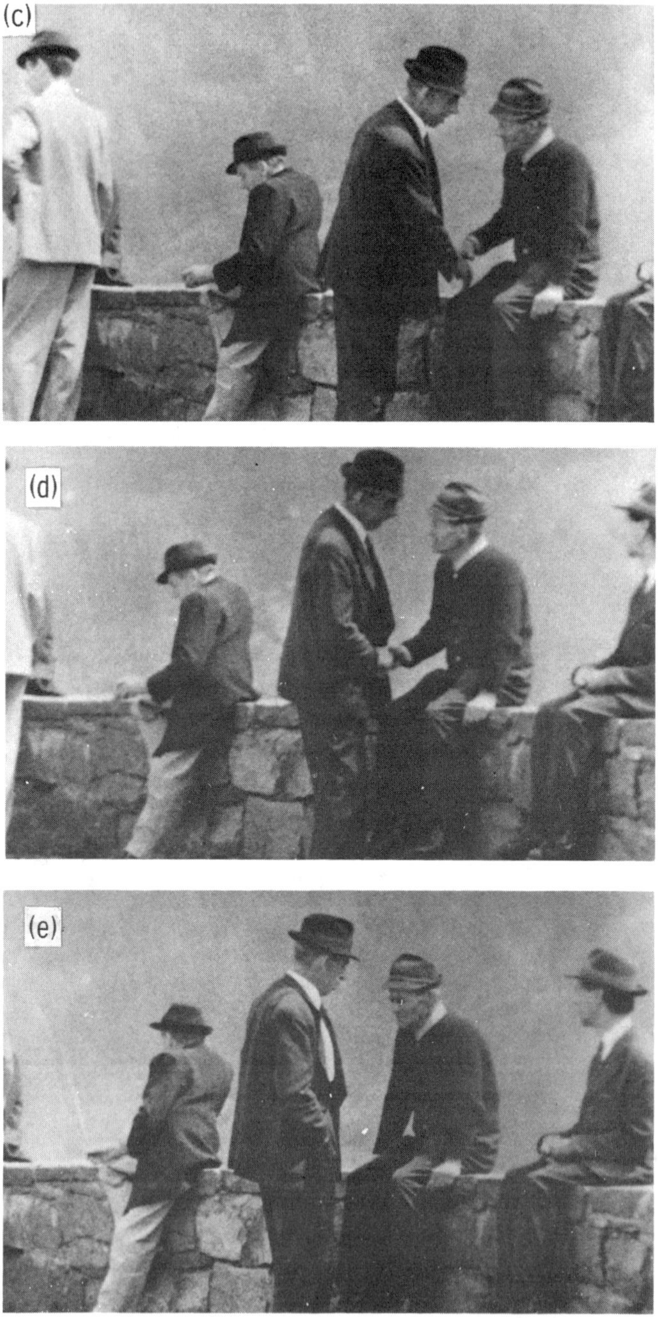

Figure 5.9 (continued)

Figure 5.9 (continued)

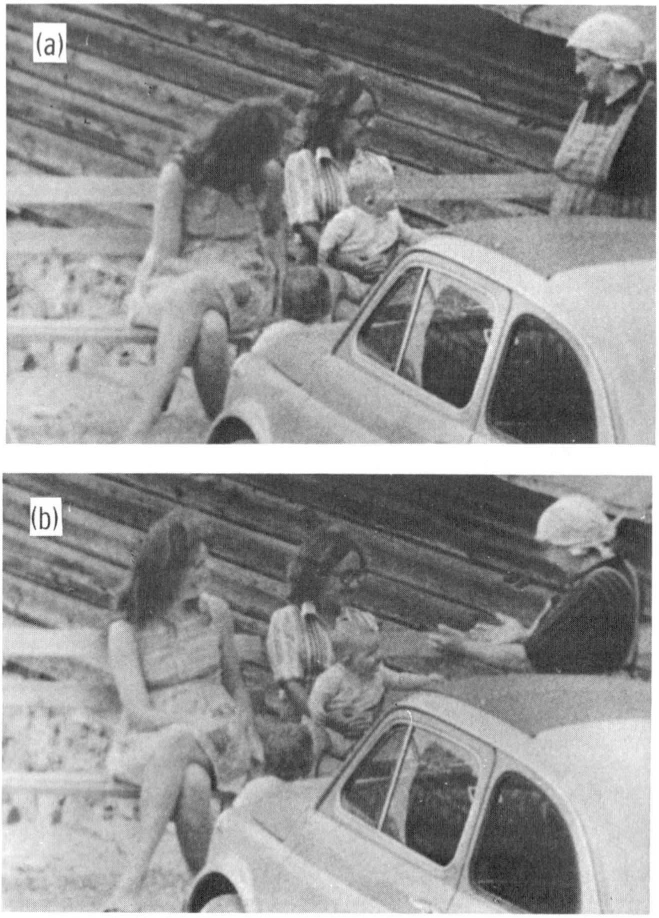

Figure 5.10 (a–h) Initiation of contact, stranger to 6-month-old infant. Note in parts b and c the infant's outstretched hands and the change of eye contact throughout the series (continued on next two pages). [Photos from 16 mm film by T. K. Pitcairn.]

Figure 5.10 (continued)

Figure 5.10 (continued)

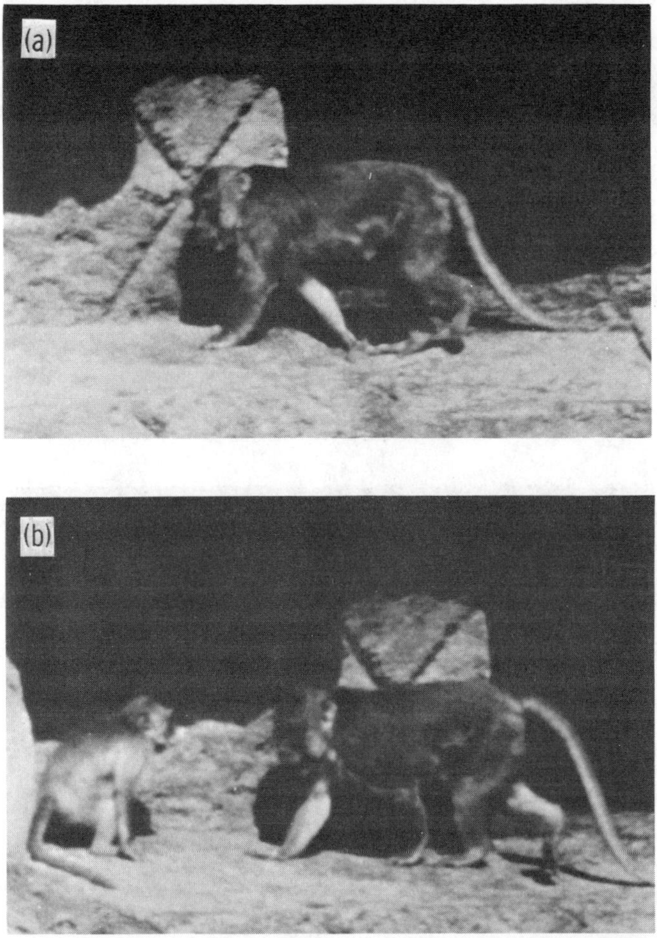

Figure 5.11 (a–e) Initiation of contact in *Macaca fascicularis,* with gaze change components throughout series (continued on next page). [Photos from 16 mm film by T. K. Pitcairn.]

Figure 5.11 (continued)

CONCLUSIONS

Thus, not only single elements of behavior can evolve, but also patterns of behavior that are contextually (situationally) rather than motivationally dependent. In fact, in any situation where a behavior is (a) stereotyped or minimally variable, at least in response to specific external factors, and (b) serves a specific function, it can be maintained in the population by some form of central programming.

This leads us then to consider the problem of variation itself. Scott (this volume) maintains that social behavior seems relatively fixed because the social structure itself is invariant. This, however, is only true when such a structure is viewed as a whole, and not when taken from the point of view of the individual within it. For example, the frequency distribution of, say, aggressive behavior or frequency of copulation is quite different among members of a monkey colony. The social structure *is* invariant, but its implications are that various individuals have fewer choices of action open to them than others. The question then is: Can such variation be included under the rubric of evolutionary open behavior? The following model makes some attempt to do this (Figure 5.12). The element of behavior produced at any one time (E) is influenced by at least six different phenomena, nested into three groups. The first pair are the structural relations of the behavior and the state (affective, or time in the sexual cycle, for example) of the individual. Then comes the variation due to social context and stream, i.e., what are the possibilities of interaction open to this particular pair, triad, etc., given the setting of the behavior and does the behavior come from a particular closed system or stream in which the possible alternatives are predetermined (as in the case of greeting behavior—here the actual form of the distance salutation and the close greeting vary according to the relationships, both in terms of time and in social terms, but the

Figure 5.12 A model of the relationships of behavioral constraints. [From Pitcairn, 1975.]

process of moving from one to the other is predetermined, i.e., has to occur). The last pair, style and channel, have not been mentioned so far, but are related to the group variations seen that are nonfunctional. The style (see, e.g., Binford, 1968) of an object or behavior is that which is laid on top of, so to speak, its primary functional determinants. One must make certain leg movements to walk from A to B, but one can walk in many different styles, and one style is generally common to a group (see Lomax, 1968). The channel is simply the biologically determined mode of communication; for example, primates mainly use the audio–visual mode of behavior transmission.

All of these factors may "explain" variation, i.e., they provide specific cues and determinants that allow a centrally organized program to be used in a relatively unfixed way. The model also serves as a central ground for biologically and culturally determined behaviors. Thus the element of style, for example, allows group identity to be maintained by a variation in a fixed behavior. However, in certain instances, for example body painting or pattern of clothes decoration, the basic activity itself may be culturally determined. The point is that these factors themselves are fixed, in that they offer performer-determined cues on which to base "decisions."

What is the importance of such work as we have discussed here? To say that a behavior is controlled either environmentally or by genetic factors says very little about its plasticity or variability. To say that a behavior is genetically fixed does not mean that one presses a button and out it comes, but rather tells us that the propensity for its production is there, and perhaps that its distribution in the population has particular features.

The importance of knowing about the control of behavior is to be able to say in what way behavior can be influenced (a) during ontogeny, (b) by modifying the environment, and (c) by learning processes. In *all* societies behavior is at some time or another controlled, from the weaning of a baby from the breast or bottle, through education, to adult life. It is surely better to know how these behaviors may be influenced, whether they are genetically controlled or not, than to adhere blindly to a program that assumes that all behavior is eternally modifiable in all directions.

REFERENCES

Ahrens, R. 1953. Beitrag zur Entwicklung des Physionomie- und Mimik-Erkennens. *Zeitschriftfur Experimentelle Angewandte Psychologie, 2,* 412–457; 599–633.

Altmann, S. A. 1967. The structure of primate social communication. In S. A. Altmann (Ed.), *Social communication among primates.* Chicago: Univ. of Chicago Press.

Ambrose, J. A. 1961. The development of the smiling response in early infancy. In B. M. Foss (Ed.), *Determinants of infant behaviour.* I. London: Methuen.

Ambrose, J. A. 1963. The age of onset of ambivalence in early infancy: Indications from the study of laughing. *Journal of Child Psychology and Psychiatry, 4,* 167–181.

Angst, W. 1974. Das Ausdrucksverhalten des Javaneraffen *Macaca tascicularis* Raffles. *Fortschritte der Verhaltensforschung, 15,* 1–90. Berlin: Paul Parey.

Ball, W., & Tronick, E. 1971. Infant responses to impending collision: Optical and real. *Science, 171,* 818–820.

Bertrand, M. 1969. The behavioural repertoire of the stumptail macaque. *Biblioteca Primatologia* II. New York and Basel: Karger.

Binford, S. R. 1968. Early Upper Pleistocene adaptations in the levant. *American Anthropologist, 70,* 707–717.

Bower, T. G. 1965. Stimulus variables determining space perception in infants. *Science, 149,* 88–89.

Bower, T. G. 1966. Slant perception and shape constancy in infants. *Science, 151,* 8328834.

Bower, T. G. 1971. The object in the world of the infant. *Scientific American, 225,* 30–38.

Bowlby, J. 1958. The nature of child's tie to its mother. *International Journal of Psychoanalysis, 34,* 350–373.

Brannigan, C. 1972. Facial expressions in a microcephalic child. Paper presented to the Association for the Study of Animal Behavior, London, June, 1972.

Chance, M. R. A. 1962. An interpretation of some agonistic postures: The role of "cut-off" acts and postures. *Symposium of the Zoological Society, London, 8,* 71–89.

Cherry, C. 1957. *On human communication—A review, a survey, and a criticism.* Cambridge, Massachusetts: MIT Press.

Eibl-Eibesfeldt, I. 1972. Similarities and differences between cultures in expressive movements. In R. A. Hinde (Ed.), *Non-verbal communication.* Cambridge: Cambridge Univ. Press.

Eibl-Eibesfeldt, I. 1973a. The expressive behaviour of the deaf-and-blind-born. In M. von Cranach & I. Vine (Eds.), *Social communication and movement.* London: Academic Press.

Eibl-Eibesfeldt, I. 1973b. Human-Ethologisches Filmarchiv. Taubblind geborenes Mädchen (Deutschland)—Ausdrucksverhalten. *Homo, 24,* 39–47.

Eibl-Eibesfeldt, I. 1973c. Human-Ethologisches Filmarchiv. Taubblind geborenes Mädchen (Deutschland)—Explorieverhalten und Spiel. *Homo, 24,* 48–49.

Eibl-Eibesfeldt, I. 1973d. Human-Ethologisches Filmarchiv. Taubblind geborenes Mädchen (Deutschland)—Soziale Interaktionen. *Homo, 24,* 49–52.

Ekman, P. 1972. Universals and cultural differences in facial expressions of emotion. In J. Cole (Ed.), *Nebraska symposium on motivation.* Lincoln, Nebraska: Univ. of Nebraska Press.

Grant, E. C., & Mackintosh, J. H. 1963. A comparison of the social postures of some laboratory rodents. *Behaviour, 21,* 246–259.

Hebb, D. O., & Thompson, W. R. 1954. The social significance of animal studies. In G. Lindzey and E. Aronson (Eds.), *Handbook of social psychology.* II. New York: Addison-Wesley.

Hooff, J. A. R. A. M. van. 1962. Facial expressions in higher primates. *Symp. Zool. Soc. London, 8,* 97–125.

Hooff, J. A. R. A. M. van. 1967. The facial displays of the Catarrhine monkeys and apes. In D. Morris (Ed.), *Primate ethology.* London: Weidenfeld and Nicholson.

Hooff, J. A. R. A. M. van. 1972. A comparative approach to the phylogeny of laughter and smiling. In R. A. Hinde (Ed.), *Non-verbal communication.* Cambridge: Cambridge Univ. Press.

Kendon, A. 1967. Some functions of gaze-direction in social interaction. *Acta Psychologica, 26,* 22–63.

Kendon, A., & Ferber, A. 1973. A description of some human greetings. In R. P. Michael & J. H. Crook (Eds.), *Comparative ecology and behaviour of primates.* London: Academic Press. Pp. 591–668.

Lomax, A. 1968. Folk song style and culture. *American Association for the Advancement of Science Publication, 88,* Washington, D.C.

McKay, D. M. 1972. Formal analysis of communicative processes. In R. A. Hinde (Ed.), *Non-verbal communication.* Cambridge: Cambridge Univ. Press.

Mackintosh, J. H. 1966. Social behaviour of laboratory mice. Doctoral dissertation, Ethology Department, University of Birmingham.

Pitcairn, T. K. 1974a. Aggression in natural groups of pongids. In R. Holloway (Ed.), *Primate aggression, territoriality, and xenophobia.* New York: Academic Press. Pp. 241–272.

Pitcairn, T. K. 1974b. Modelle zur Erforschung nonverbalen Verhaltens. Paper presented to the 29th Kongress der Deutschen Gesellschaft für Psychologie. Salzburg, September, 1974.

Pitcairn, T. K. 1975. Attention and social structure in *Macaca fascicularis*. In M. R. A. Chance & R. R. Larsen (Eds.), *The structure of social attention*. London: Wiley.

Pitcairn, T. K., & Grant, E. C. 1972. A dyadic analysis of human behaviour. Paper presented to Association for the Study of Animal Behaviour, London, November, 1972.

Robson, K. S. 1967. The role of eye-to-eye contact in maternal-infant attachment. *Journal of Child Psychology and Psychiatry, 8,* 13–25.

Spitz, R. A., & Wolf, K. M. 1946. The smiling response: A contribution to the ontogenesis of social relations. *Genetic Psychology Monographs, 34,* 57–125.

Chapter

6

THE ONTOGENY AND PHYLOGENY OF SOCIAL INTERACTIONS[1]

ROBERT B. CAIRNS
University of North Carolina at Chapel Hill

Embryos undergo development; ancestors have undergone evolution, but in their day they also were the products of development.

—Gavin de Beer (1958, p. 1)

The centennial of Charles Darwin's *Expression of the Emotions* has come and gone and the relationship between behavioral ontogeny and phylogeny remains unresolved. The problems are not merely definitional, although the area resembles a semantic mine field (Lehrman, 1970). Nor are they only problems of criteria, even though the ground rules for classifying events as behavioral homologues are

[1] This work was supported in part by NICHHD grant HD 086464-02.

subjective and tautological (Atz, 1970; Gottlieb, 1970). Both problems are surface features, it appears, of what are basic differences in perspective on what constitutes an adequate explanation for social development. The differences in perspective, in turn, reflect two fundamentally different orientations to the ontogenetic–phyletic relationship.

TWO VIEWS OF THE ONTOGENY–PHYLOGENY RELATIONSHIP

One orientation would have it that phylogeny is the key to explaining development (Lorenz, 1974; Tinbergen, 1974). So aggressive drives and expressions arise inevitably in the descendants (including humans) because of the historical survival functions of aggression (i.e., Eibl-Eibesfeldt, 1967; Lorenz, 1966). And social attachment behaviors come about because they are part of a "genetic blueprint" for protection from danger (Ainsworth, 1973; Bowlby, 1958, 1969, 1973). From this orientation, a major task for research is to demonstrate that certain attachment behaviors—such as infant protest-crying and mutual visual regard—are indeed homologous to those in the ancestral species. Once the cross-taxon similarity has been demonstrated, a major explanatory problem is solved. Accordingly, phylogeny has been assumed to play a causal role in human development. This, the ancestor's view of the ontogeny–phylogeny relationship, can be summarized by the expression:

$$\text{PHYLOGENY} \longrightarrow \text{ONTOGENY}$$

This proposal is a familiar one, having been adopted by many developmental psychologists as the appropriate way to view the linkage between phyletic relationships and the development of behavior.[2] As plausible as the argument may be, a critical examination suggests that there is a significant gap in the logic of the

[2] This statement is not simply a resurrection of the Haeckelian biogenetic law, summarized in the trenchant phrase: "ontogeny recapitulates phylogeny." The recapitulation argument fell in the face of morphological/anatomical evidence that development had its own distinctive features and

proposal. The problem is that it fails to specify *how* phyletic relationships among ancestral species can exercise a direct effect upon behavioral development in the young of the successors. In considering this matter, de Beer (1958) has offered the counter-proposal that the relationship is the other way around—that phylogeny can play no causal role in determining morphological development, but that modifications that occur in successive ontogenies can bring about changes in a phyletic series. Here the focus is on the developing organism and what it can tell us about evolution. Further, de Beer (1958) observes that "Phylogeny is provisionally to be regarded as a series of adult forms, which are disconnected and causally unrelated to one another, each adult form being the result of an ontogeny which differs from the previous one [p. 171]." Successive ontogenies are related to one another by the transmission of internal factors from fertilized egg to fertilized egg. And modifications in these internal factors during ontogeny bring about modifications in the adult forms (i.e., phylogeny). This is the embryo's view, or

$$\Delta \text{ ONTOGENY} \longrightarrow \Delta \text{ PHYLOGENY}$$

Accordingly, changes in successive ontogenies lead to modifications in phylogeny.

Precisely what kinds of ontogenetic variation might effect differences in the phyletic series? The pioneering work of D'Arcy Thompson (1917) and Garstang (1922) on the processes of growth in evolution provides an essential clue. These investigators found that major changes in anatomical structure and neurological organization come about by a quantitative alteration of the rate of maturation of particular characteristics. *Heterochrony,* the term that describes these changes in timing, has proven to be a powerful concept in accounting for morphological change in a phyletic series.

One way that the differential timing mechanism can operate is through the retention into the maturity of the descendants of certain

---

mechanisms that could not be reduced to a speculative analogue to evolution. In contrast, the contemporary arguments on behavioral evolution would have it that basic behaviors and motives have been retained in descendants because of their adaptive functions in phylogeny.

foetal or youthful features of the ancestral species. Some of the morphological characters of humans reflect a retardation of rates of growth relative to the lower primates and great apes. This includes such characteristics as the delay in closing the sutures of the skull, without which the continued development of the brain might not have been possible. Other instances include the retention into the adult human form of certain foetal characters of nonprimate mammals, such as the position of the skull relative to the trunk (the cranial flexure), which has been speculated to be responsible for the erect posture of man (de Beer, 1958). Such a retention of youthful characters of the ancestor has been called a *neotony.* An alternative process, *acceleration,* involves the appearance of a mature character of the ancestor in the youthful stages of the descendant.

The differences in perspective on the ontogeny–phylogeny relationships are rather important ones, I believe. If the embryo's view of evolution is taken seriously, it may provide some basis for bridging the gap between seemingly irreconcilable views on social development.

SOME INSTANCES OF HETEROCHRONIES IN BEHAVIOR

But first there is the problem of generalization. Psychologists have gotten into no little difficulty by plagiarizing concepts from other fields. Words like drive, valence, and homology come to mind. Terms that have a precise meaning in their original context tend to become blurred in the transition. And that brings us to our present problem. The concepts of heterochrony and its corollaries, acceleration and neotony, are essentially morphological constructs. Can they be applied to issues of social development in any nontrivial way?

There is some evidence that they can. A critical overview of the uses of the concept of neotony indicates that it has occasionally been employed in the discussion of behavioral phenomena (Bekoff, 1972; Geist, 1971; Kuttner, 1960; Mason, 1968). Geist (1971), for example, considers the physique and behavior of Rocky Mountain bighorn rams to be a neotonization of adult rams of more primitive races. In particular, aggressive interchanges among Rocky Mountain rams are similar in form and type to those found in juvenile Stone's sheep (a thinhorn sheep of British Columbia that resembles the common ancestor). Similarly, structural–skeletal similarities are found between the adult bighorn ram and the juvenile Stone's sheep male. On a broader scale, William Mason (1968) has cogently argued that neotony is a general primate characteristic and that

> man has gone farthest in this evolutionary venture—a venture in which the reliability and efficiency of instinctive patterns have been sacrificed to achieve the behavioral plasticity and the liberation of psychic energies that play so large a part in human frailties and human achievements [Mason, 1968, p. 112].

Mason's proposal—that the neotonous development of humans is characterized by great fluidity and less dependence on instinctive bases than nonhuman development—deserves serious attention.

In addition to instances where the concept of neotony has been explicitly applied to behavior, there are occasions where the idea has been used even though the term has not. One especially interesting example of the implicit usage concerns the development of strain differences in audiogenic seizure susceptibility. The phenomenon itself is a striking one: Some strains of mice enter a convulsive state, often fatal, if exposed to a loud ringing bell. Tested at a particular time—at 21 days of age (weaning)—strains can be reliably classified in terms of their seizure susceptibility. Certain inbred strains will virtually always convulse; other strains, rarely. Of particular relevance to our present discussion was the discovery that a neat classification of "susceptible" and "nonsusceptible" strains breaks down when the age of testing is varied (Fuller & Sjursen, 1967). Some strains classified as nonsusceptible at 21 days of age turned out to be highly susceptible when tested just 1 week later (28 days of age).

Why should 1 week make a difference? In considering this problem, Fuller and Sjursen (1967) proposed that the increased susceptibility may reflect a quantitative imbalance that occurs in development of normal physiological systems, as opposed to a unitary neurological deficit. They argue that the variability among strains in seizure susceptibility may be due to genetically controlled differences in the rate of maturation of the component physiological processes. The imbalance responsible for higher susceptibility would then occur at different times for different strains. In brief, heterochrony in physiological development can lead to marked strain differences in behavioral responsiveness.

Insofar as social development is also paced by physiological processes, it seems reasonable to expect that differences among strains in social behavior may themselves reflect heterochrony in maturation. A phenomenon that we encountered recently provides

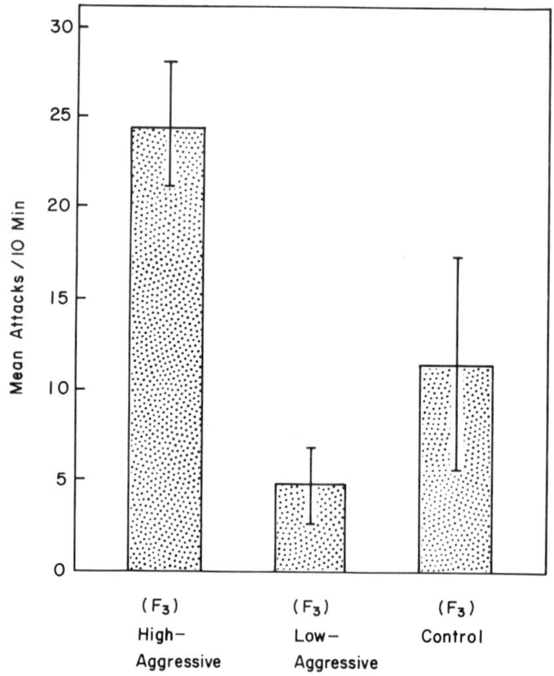

Figure 6.1 Differences among the male mice of two selectively bred lines in terms of mean attacks are highly reliable ($p < .001$) in the F_3 generation. The animals were tested at 64 days of age ($\pm$ 3 days) after having been maintained in isolation since weaning. The selection experiment began with a foundation stock of 16 male and 16 female outbred ICR mice. (Standard error of mean shown by bars.)

some support for that expectation. Over the past $2\frac{1}{2}$ years, we have studied how selective breeding can influence aggressive behavior in mice. The selection program was itself reasonably successful, as Figure 6.1 shows. By the third filial generation—using standard rearing and cross-fostering techniques—the animals were reliably segregated. The outcome is consistent with the work of the Lagerspetz at Turku (Lagerspetz & Lagerspetz, 1968; Lagerspetz, Tirri, & Lagerspetz, 1971). Moreover, a correlated behavioral characteristic—which may act as a primary behavioral mediator for differences in attack—was identified. Direct behavior analyses of the reciprocal interchanges prior to attacks indicated that the animals in the two different lines reacted differentially to social stimulation. When sniffed by another animal, those in the low-aggressive or nonaggressive line tended to freeze or become rigidly immobile prior to showing a reflexive kick (see Figure 6.2). As a group, they tended not to initiate interchanges and to become immobile when contacted. For these animals, there was little initiation of any sort, including those kinds of behaviors that might escalate into attacks. Animals in the high-aggressive line tended to explore actively and respond to their test partners with vigorous counter-actions; they rarely froze. The heightened activity–reactivity was typically a precursor to the occurrence of attacks.

Although the selective breeding work looked successful, two features of the experiments bothered me. First, the studies were almost *too* successful. Clear differences were obtained as early as the first and second generations. That seemed unusually soon. Second, and perhaps more important, the phenomenon was directly linked to the age of the animal at the time of testing. Young animals—those tested at 64 days of age, the time set for critical selection tests—showed the biggest differences as a function of breeding line. When they were retested later, at about 6 months of age, the differences between the groups diminished. This is shown in Figure 6.3. The so-called nonaggressive animals were highly likely to attack their partners, when tested at 6 months of age. It was of interest that the 6-month-old animals were also less likely to freeze or become immobile on stimulation. Rather, dyadic stimulation was followed

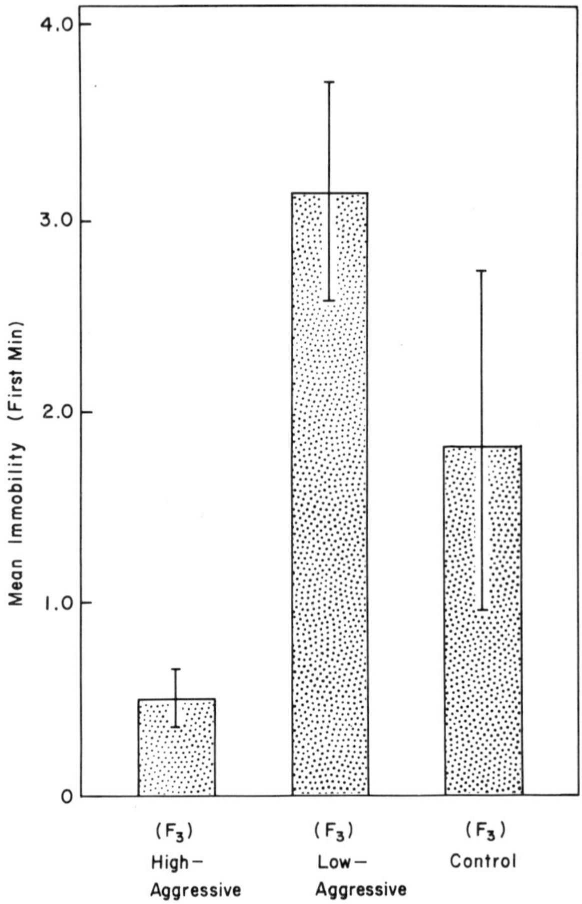

Figure 6.2 The mean frequency of immobility responses observed among the F_3 males shown in Figure 6.1. The differences in immobility as a function of selection line were highly reliable ($p < .001$). Immobility was observed when the animal was stimulated (e.g., nosed or sniffed) by another animal in the test compartment. The reaction is presumed to be one of the factors that is responsible for the relatively lower level of fighting observed among the "low-aggressive" animals. (Standard error of means shown by bars.)

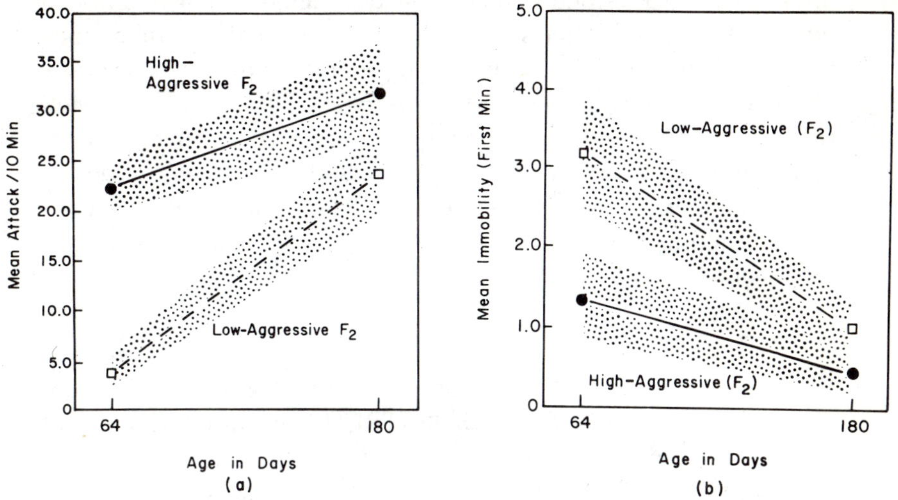

Figure 6.3 The differences between the high- and low-aggressive males (F_2) in the mean number of attacks (a) and mean immobility (b) as a function of chronological age at the time of testing. In attacks (a), the difference between lines is highly reliable at Day 64, but the difference is not significant when tested at Day 180. A similar interaction between age and immobility is observed (b), with significant differences obtained only at Day 64. (Standard error of means shown in shaded areas.) Comparable effects were obtained when the animals were tested in a cross-sectional design (data not shown).

rapidly by counter-exploration and investigation, including the vigorous type that led to fighting. For the F_2 generation, the phenomenon was age-correlated.

These results suggest that the changes produced were more general than the discrete social activity (fighting) assessed. In the process of selecting for fighting, we had apparently varied the maturation rate of the physiological systems related to arousal–reactivity. The behavior of animals in the nonaggressive line became increasingly more similar to that of animals in the aggressive line as they (the "nonaggressive" line) grew older. Whether differences in the maturation rate of component physiological systems will be adequate to explain the behavioral differences in the lines remains to be determined. Our best guess is that it is not the whole story.[3] Nonetheless, it is significant that multiple social behavior differences are influenced by a modest shift in the timing of physiological development. More generally, in assessing phyletic similarities and differences, one must be as sensitive to *when* a test is conducted as to *what* is being assessed. Moreover, a genetic sledgehammer is not always required to demonstrate widespread social outcomes; a delicate retiming may be sufficient to produce significant changes.

PHYLOGENY AND THE DEVELOPMENT OF SOCIAL INTERCHANGES IN CHILDREN

Now we will shift to a related but considerably more speculative matter: behavioral ontogenesis and the evolution of positive social interchanges in children. Much has been written recently about the phyletic contributions to the development of the mother–infant relationship in mammals. John Bowlby (1958, 1969, 1973), in particular, deserves credit for having vigorously argued the importance of the evolutionary perspective in explaining why and how the relationship comes about in humans. But the evolutionary perspective has been limited, for the most, to statements of the phylogeny ⟶ ontogeny type, or speculations about what phylogeny can tell us about social ontogeny. If the embryological perspective has merit, there should be another side of the story, one that focuses on how changes in ontogeny may have contributed to differences in behavioral evolution.

From the embryological perspective, the unraveling of the relationship between the ontogeny of children and those of nonhuman species requires at least two steps. First, one must identify what the common major features of social development are in children and in the young of related species. Second, one must assess whether

[3] Equally important, residual differences remain between lines in (a) attack latency, (b) sexual potency, and (c) attack vigor, in the late test series. These results will be summarized in detail in a research report (in preparation).

the developmental timing of these features in human development is accelerated or retarded with respect to the timing of the features in nonhuman mammals. There are pitfalls at both levels, and the construction of even a tentative map of behavioral evolution across species that show multiple differences is a risky business. And since the ancestors cannot be observed directly, we must use information about species which themselves have continued to evolve. What follows is a speculative answer to the question of whether the social development of children bears a neotonous or accelerated relationship to the ontogenies of our predecessors.

For our present purposes, three features of child social development are of particular interest: early unreliability, flexibility, and plurality.

In the initial year of life, the infant begins to distinguish among persons in his immediate environment. The age of onset of discrimination depends on the response modality assessed. Prior to 6 months of age, the child can and does register recognition and discrimination by changes in facial expressions, differential orientation, and smiling. By the end of the first year, the discrimination–preferences become sharpened, and most infants vigorously cry on some occasions when they are separated from other persons, notably the father or mother. The child will, under particular conditions, selectively approach others. The infant's discriminations are age-related, reflecting the maturation of basic perceptual and response processes.

The last point is an important one. Befitting the immature and rapidly changing status of the infant, variations between infants in their social behavior are not highly stable from one time to another. Individual differences in terms of "crying–separation" and selective approach behaviors turn out to be notoriously unreliable. For children 1 and 2 years of age, one can scarcely predict better than chance over a 3-month interval which infants are going to be more "socially attached" than others (Coates, Anderson, & Hartup, 1972; Lewis, 1971; Maccoby & Feldman, 1972; Masters & Wellman, 1974; Stayton & Ainsworth, 1973). Predictions over a 6-month period are negligible. Such instability presents a problem for infant typologies and the extrapolation of early maternal–infant relationships to life-long adjustment patterns (Ainsworth, 1973; Bowlby, 1973). However, the findings do fit rather well with the view that the child's early response patterns and cognitive capabilities undergo rapid changes that in turn promote variability in the mother, the offspring, and their relationship (Cairns, 1972).

A second and related feature concerns *adaptation* and *malleability*. In fact, the human infant remains highly adaptable to new social patterns and relationships. In this regard, D. E. Fleener (1967) found that some primary "attachment" behaviors could be produced within a very short period—7½ hours. The year-old child interacted with a caretaker for a short period for each of 3 days. According to which caretaker was chosen, the child showed a strong preference for that individual. The intensity and strength was the same as observed in the child's relation to his natural mother—but the difference is that the caretaker had interacted for only 2½ hours on each of 3 days!

Fleener's work was recently extended in a doctoral dissertation by L. C. Monahan (1975). Monahan explored how children in early and late infancy react differentially to a stranger as opposed to the child's own mother. Among other things, she found that most of the infants responded quite positively to an unfamiliar adult in the home, *if* the person responded like a real human being. Indeed, the infants demonstrated somewhat *more* positive responses to the stranger than to their own mothers. They smiled more, laughed more, and were generally more socially responsive to the stranger. But what must also be noted is that the stranger was pretty nice as well; she was most active, vocal, and smiling. We should maybe call her the "other woman" instead of a stranger. Monahan took the next step and developed dyadic interchange analyses, teasing out the conditional relations between the occurrence of the several acts of the adult and child. She found that smiling in the infant was conditional upon particular acts of the adult, and that the "other woman" performed more of these acts than the mother. Mutual visual regard was one such act sequence; adult orientation and talking was another. Conversely, crying seemed not to be particularly elicited by the stranger or mother, and could be differentially controlled by particular acts of either person. In both instances, crying could be elicited by an abrupt termination of the relationship, such as leaving the room. This was no different for the mother than for the "other woman."

But are not children supposed to fear strangers? Some do, and Monahan's work was no exception. However, just 3 of the 24 children tested showed extended fearful crying in Monahan's work, a finding that replicates the reports of Corter (1973) and Rheingold and Eckerman (1973). Why should some investigators find stranger anxiety and fear and others find adaptation? At least part of the difference lies in the nature of the other woman's behavior. In those cases where children are most fearful, they have good reason to be. The stranger is not only unfamiliar but he is forbidding, nonreactive, and expressionless. The "fear" may itself be elicited by the distinctive nonreciprocal characteristics of the individual, not merely his unfamiliarity. In any case, reasonably synchronous or reactive behaviors on the part of the adult tend to elicit comparable and reciprocal actions by the child. When these occur in the relationship, they provide the behavioral foundations for a reasonably compatible and mutually acceptable pattern of interchange.

Overall, these studies indicate that young children are not insulated from other people by an invisible "attachment" shield erected around them by the mother. To the contrary, the adaptation processes are ongoing. The same kinds of dynamic interchange events that make for the mother–child interactions are present in other relations as well. All of which make a lot of sense when one takes a broader look at the multiple social relationships in which children eventually become involved.

That brings us to a third major character of social patterns: the relational and contextual specificity of the child's interaction patterns. By age 4 and possibly earlier, the child not only forms interchange patterns that are distinctive to particular other persons, including same-age children and adults, but these patterns vary directly as a function of the context in which they occur. One of the interesting findings to emerge from the detailed observation of preschool behavior patterns are their contextual and relational relativity (Cairns, Sherman, Holmberg, Buck, & Patterson, in preparation; Garvey & Hogan, 1973; McGrew, 1972). The phenomena reflect an adaptive plurality in social interchange patterns. Such multiplicity has sometimes been confused with the notion of reversibility of attachment. That confusion is unfortunate, because it implies that the mother–child relationship itself was somehow extinguished or drastically modified. Such a view is too narrow for the facts. The child's relationship with the mother is ongoing at the same time that other relationships are established. Nor is the concept of attachment very helpful in accounting for how multiple relations are formed and maintained, each appropriate to a particular setting or pattern of stimulus conditions. The same kinds of process that make for adaptation to the mother and her distinctive characteristics appear to provide the tools for the child's adaptation to other persons. This capability is not extinguished or outgrown at infancy; rather it has just gotten underway.

SOCIAL DEVELOPMENT IN NONHUMAN YOUNG

Now we can return to the major concern here, namely, how are these characteristics of the child related to those found in species that evolved from common ancestors? The neotony proposal is of special interest in this regard. In morphological development, humans show a retention of some of the youthful or fetal physical characters of primates and other nonhuman mammals. Some of these have already been mentioned: the delayed closure of the skull sutures, the retention of the cranial flexure into maturity, and the persistence of a state of relative hairlessness. Certain of these morphological features are critical, and others peripheral, to man's distinctive adaptations.

Consider now how the three major characteristics of child social development—instability, malleability, and multiplicity—appear in nonhuman social development.

Formation and Reliability

Although the establishment of the mother–infant relationship in nonhuman mammals is well documented, the stability of individual differences in the relationship is not. In the early stages of life, it is the mammalian mother, not her offspring, that bears the major responsibility for the preserving relationship. Neonatal sheep, monkeys, dogs, cats, and mice are relatively indiscriminative about whom they will approach and suckle (Cairns, 1966, 1972; Harper, 1971; Rheingold, 1963). Among mammalian species that form strong and distinctive social bonds (e.g., sheep, many primate species), the unreliability gives way early to stability, where there is a shared responsibility between the mother and her offspring for the maintenance of the relationship. Not all mammalian young, however, develop strong and distinctive social relationships. Mice and rats, for instance, do not show much evidence of strong individual social bonds even at maturity. Variations in the exclusiveness and intensity of the maternal–infant relationship have been linked to such characteristics as the number of offspring (one versus multiple), species-typical ecological demands, and the social structure of the group. However, among those species that are characterized by exclusive maternal–infant social preferences, the neonatal period appears to be the one where there is the greatest instability and flexibility.

PLASTICITY OF INTERCHANGES

Information on the plasticity of social behavior is limited, primarily because developmental researchers have been reluctant to explore how the effects of early experience can be modified, extended, or reversed. One can only speculate about the reason for the reluctance. One possibility is that developmental research strategies have been determined by acceptance of the primacy hypothesis—that what happens first is necessarily more lasting than what happens later. Whatever the reason, until recently scant attention has been given to the processes that may mediate and account for the persistence or reversal of early experiences (Simmel, 1973).

Fortunately, some recent influential studies have been addressed to the problem of social behavior plasticity. One set has been concerned with the persistence and change in imprinting effects in precocial birds; other investigations have explored the reversal of early isolation effects in mammals. The upshot of the imprinting work has been to show that not only are "primary" imprinting effects subject to modification, but that the changes can take place quite rapidly (Einsiedel, 1973; Salzen & Meyer, 1967; Zajonc, Reimer, & Hausser, 1973). Working with domestic chickens, Salzen and Meyer (1967) found that the simple manipulation of interchanging stimulus objects (i.e., substituting an unfamiliar object for the initially imprinted one) eventually produces a shift in the young chick's preference. After 72 hours of experience with the "new" object, the chicks clearly preferred it over the one to which they had been originally imprinted. In a replication and extension of Salzen and Meyer (1967), Einsiedel (1973) found that there was a direct correlation between the length of exposure to the "new" object and the strength of its preference for that object: the longer the time spent with the "recent" object, the greater the tendency to approach that object in a two-object test. Moreover, significant shifts began to appear after only 5 hours in the "interchange" condition.

Are these instances of the reversibility of imprinting? The outcomes of the simultaneous test suggest that they are. However, when Einsiedel (1973) presented the two objects sequentially—one at a time—the animals approached each of them in turn. Einsiedel concluded that the young chicks become adapted to both the "initial" and "recent" imprinting objects, and that the apparent "reversal" was a feature of the simultaneous test procedures, not of the animal. Two preferences could co-exist, if the test permitted it. In effect, the range of stimuli to which the young animal responded was *extended* instead of reversed.

Information on the plasticity of social patterns in young mammals is also limited, but what we have is likewise significant. Since the classic instances of social nonreversibility in mammals were reported in studies of isolation-reared dogs and primates, it is of interest that recent studies of these species have forced a re-evaluation of the permanence of the early social deprivation effects (Fuller, 1967; Suomi & Harlow, 1972).

Consider the case of the alienation syndrome produced by isolation rearing in the rhesus monkey. Early accounts emphasized the refractory nature of isolation effects on subsequent maternal, sexual, and social behaviors. But more recent reports indicate that isolate-reared monkeys can, under some conditions, develop entirely acceptable patterns of social interchange (Mason & Kenny, 1974; Suomi & Harlow, 1972). Why the apparent discrepancy in findings? Suomi and Harlow speculate that the problem was the nature of the therapist. In their successful treatment, the "therapist" was a young, nonthreatening female monkey. The young animals tended to elicit reciprocal play, and, generally, benign patterns of mutual interchange. When isolated animals are placed with other same-age males, isolated or not, the interactions rapidly escalate to mutual fighting and mayhem. They proposed that the key was the nature of the receiprocal feedback, precisely what one would expect on the basis of a dyadic analysis (Cairns, 1972).

Further, Mason and Kenny (1974) have shown that the "partner" need not be of the opposite sex, nor need it be a primate. Indeed, they demonstrated that young isolated monkeys could be made more socially responsive merely by maintaining them with a friendly dog. The changes in the behavior of the previously "alienated" monkey were rapid and dramatic. Mason and Kenny indicate that

> Initially, most monkeys reacted to the dogs with fear, expressed in grimaces, distress vocalizations, crouching, and withdrawal. These behaviors usually disappeared quickly, however. All but one of the eight subjects approached within 2 hours (5 within 30 minutes), and all approached within 6 hours [Mason & Kenny, 1974, p. 1210].

Reports of the malleability of young dogs and sheep yield results that are not inconsistent with this picture of rapid malleability (Fuller, 1967). In particular, young puppies will rapidly adopt species-atypical preferences, if the conditions of rearing and housing are appropriately arranged (Cairns & Werboff, 1967; Kuo, 1960). Again, the process itself can be timed in minutes and hours as opposed to days and weeks.

Relationship Plurality

Information is least adequate on the third characteristic, the multiplicity of relationships. At an anecdotal level, the answers seem clear. Observations of juvenile monkeys and dogs indicate that there is considerable differentiation across relationships. Yet few investigators have gone beyond the clinical observation of the phenomenon. Nowlis (1941) did. He suggested, in effect, that adolescent chimpanzees have a companion for all seasons: one for play, another for roughhousing, and another for sleeping. Beach (1969) has plotted the multiple preferences that obtain in dogs. Unfortunately, our recent concern for plotting key "critical" relationships, such as imprinting and attachment, may have obscured the study of social interactions that may be equally important in species identification and propagation. In any case, there is at least some reason to believe that multiple relationships can co-exist in nonhuman adolescents and adults, even though little is known about their ontogeny or contextual and relational controls. That seems to be worthy of further investigation.

NEOTONY OR ACCELERATION?

Returning now to the issue that inspired this overview: Does the social behavior of young humans bear a neotonous relationship to nonhuman mammals? In this regard, one of the hallmarks of the behavioral organization of the mammalian neonate as opposed to that of the adult is the high degree of malleability and flexibility of its social behavior patterns. As Kummer (1971) has rightly observed,

> In contrast to the highly specialized but rigid skills of lower vertebrates, then, primates have a potential for learning broad sets of tasks which neither they nor their ancestors encountered in this particular form. This flexibility, and not a specialized but genetically fixed skill, prepared the way for culture [Kummer, 1971, p. 147].

Kummer goes on to speculate that flexibility in social relationships may have preceded and predisposed the ancestors of man to develop flexibility in the use of technical, nonsocial tools.

Perhaps this basic social and behavioral flexibility—as apart from special homologues or other fixed social patterns—is the character of social development that is neotonized. Two of the three features of child social development discussed would seem to be consistent with such a generalization. The instability of individual differences and the high malleability of social interchanges throughout early childhood appear to be found in the very early social adaptations of all young mammals. In children, however, both properties of social relationships persist beyond the neonatal period.

The third feature—the multiplicity of relationships—does not appear to be a neotonization in children of a basic characteristic found in the youthful behavioral adaptation of ancestors. On the contrary, the child's capabilities for multiple relationships seem to be *accelerated* relative to those of "more primitive" species. Along with the acceleration in multiple interchange formation, there are correlated capabilities for symbolic communication and classification of persons, objects, and places. Hence, in the child one finds superimposed on a relatively immature physical structure an advanced development of communication–discrimination processes.

CONCLUDING COMMENTS

What might be concluded from this brief comparative overview of the relations between ontogeny and phylogeny in social development? The evidence supporting the claim that there should be a re-examination of the directionality of the ontogeny–phylogeny relationship is most compelling where differences between the behaviors of closely related species or breeding lines are considered. Geist's (1971) interesting analysis of the relationship between races of sheep is an example of how the study of behavior evolution can proceed with existing species. Even more precise have been attempts to study relatively isolated characters in genetically controlled lines, such as has been done in the analysis of audiogenic seizures and the development of behavioral propensities associated with fighting in mice. It seems not unreasonable to expect that other socially relevant behaviors might be similarly traced to variations in the rate of development of basic physiological systems.

More recently, Gottlieb (1971) has extended the epigenetic argument in the context of species-identification processes in birds. Specifically, he makes the important point that the regulation of the timing of developmental events is jointly determined by organismic (including genetic) and sensory stimulative factors. He writes,

> In the normal course of development, the manifest changes and improvements in species-specific perception do not represent merely the unfolding of a fixed or predetermined organic substrate independent of normally occurring, sensory stimulation. With respect to the evolution of species-specific perception, natural selection would seem to have involved a selection for the *entire developmental manifold*, including both the organic and normally occurring stimulative features of ontogeny [Gottlieb, 1971, p. 156; emphasis added].

Less support can be claimed for the specific hypothesis that human social development is a neotony of primate social development. Some features of the social behavior of children—notably their extended period of physical immaturity and a correlated flexibility and instability of early interchange patterns—fit the neotony hypothesis. However, other features of the child's communicative development do not. These include the accelerated use of language and symbolic communication, along with advanced memory storage and recall capabilities. Considered in overview, both major types of heterochronies—neotony *and* acceleration—seem to be involved in the multiple differences between human and related primate forms. An appreciation of the possibility that variations in the timing of gene expression are not merely neotonous may help to resolve some of the dilemmas in the "neotony" account of the evolution of human social behaviors.[4]

One further comment seems in order on the relationship of this general proposal to recent advances in the study of the molecular bases for evolution. Biochemical analyses of the DNA base se-

[4] One other implication of the neotonization proposal should be underscored. Contrary to a "primacy" assumption, the earliest relationships of the young child may be less "foundational" than are later ones in understanding eventual social adaptation in adolescence and at maturity. Infancy relationships, relative to those in later childhood, are clearly less stable, more easily modified, more readily extended, and less often recalled. A revised view of human social development indicates that attention might well shift to stages where there is a greater likelihood for the child's interchanges to be replicated, generalized, and consolidated.

quences of humans and chimpanzees indicate that there are great similarities in macromolecular structure. Using standard biochemical methods for estimating genetic distance, King and Wilson (1975) report the remarkable finding that the human polypeptide is "more than 99 percent identical to its chimpanzee counterpart [p. 114–115]." In other words, the genes of humans and chimpanzees are as similar as are those of sibling species in other organisms (i.e., sibling species are almost identical in terms of morphology). Why, then, the vast differences between the behavior and anatomy of humans and chimpanzees? King and Wilson (1975) proposed, in an argument parallel to the one offered in this paper, that the major differences result from mutations that influence the expression of genes. Accordingly, they speculated "small differences in the time of activation or in the level of activity of a single gene could in principle influence considerably the systems controlling embryonic development [p. 114]."

Particularly crucial for the King and Wilson (1975) proposal would be the demonstration of differences between chimpanzees and humans in the timing of gene expression during development of the central nervous system. However, critical evidence on the biochemical bases of heterochrony awaits the discovery of more sensitive techniques for detecting and controlling regulatory differences at the molecular level in mammals.

Finally, the limits of an evolutionary analysis of behavioral development—regardless of directionality—should be underscored. Although speculations about the species-survival functions of social behaviors may help us to view human development in a broader biological perspective, they cannot substitute for explanations of how the interactions develop. Precise answers to the critical questions of social development—on the controls of stability, malleability, and multiplicity—require precise information about development of the particular species involved. The primary task is to identify the ontogenetic mechanisms that are operative in a given species by experimental, observational, and comparative procedures. This need remains as true for human beings as for nonhuman mammals. Evolutionary speculations have proceeded too far on the assumption that we already know the basic characteristics of communication and social development across species. We do not. Relative to other areas of biological science, work on social development remains in a primitive stage for both humans and animals.

ACKNOWLEDGMENTS

Many of the ideas in this essay were discussed in animal behavior and comparative-developmental seminars at Indiana University and the University of North Carolina. I owe a special debt to two colleagues who collaborated in these ventures and from whom I learned much: Donald D. Jensen at Indiana, and Gilbert Gottlieb at UNC.

REFERENCES

Ainsworth, M. D. S. 1973. The development of mother–infant attachment. In B. M. Caldwell & H. N. Ricciuti (Eds.), *Review of child development research.* Vol. 3. Chicago: Univ. of Chicago Press.

Atz, J. W. 1970. The application of the idea of homology to behavior. In L. R. Aronson, E. Tobach, D. S. Lehrman, & J. S. Rosenblatt (Eds.), *Development and evolution of behavior: Essays in memory of T. C. Schnierla.* San Francisco: Freeman. Pp. 53–74.

Beach, F. A. 1969. Locks and beagles. *American Psychologist, 24,* 971–989.

Bekoff, M. 1972. The development of social interaction, play, and metacommunication in mammals: An ethological perspective. *Quarterly Review of Biology, 47,* 412–434.

Bowlby, J. 1958. The nature of the child's tie to his mother. *International Journal of Psychoanalysis, 39,* 350–373.

Bowlby, J. 1969. *Attachment and loss.* Vol. 1. *Attachment.* New York: Basic Books.

Bowlby, J. 1973. *Attachment and loss.* Vol. 2. *Separation.* New York: Basic Books.

Cairns, R. B. 1966. Attachment behavior of mammals. *Psychological Review, 73,* 409–426.

Cairns, R. B. 1972. Attachment and dependency: A psychobiological and social-learning synthesis. In J. L. Gewirtz (Ed.), *Attachment and dependency.* New York: Wiley. Pp. 29–80.

Cairns, R. B., Sherman, S. J., Holmberg, M. C., Buck, N. L., & Patterson, D. A. In preparation. Interchanges in the pre-school: Stability, multiplicity, and contextual relativity.

Cairns, R. B., & Werboff, J. 1967. Behavior development in the dog: An interspecific analysis. *Science, 158,* 1070–1072.

Coates, R., Anderson, E. P., & Hartup, W. W. 1972. The stability of attachment behaviors in human infants. *Developmental Psychology, 6,* 231–237.

Corter, D. M. 1973. A comparison of the mother's and stranger's control over the behavior of infants. *Child Development, 44,* 705–713.

de Beer, G. 1958. *Embryos and ancestors.* (3rd ed.) London: Oxford Univ. Press.

Eibl-Eibesfeldt, I. 1967. Ontogenetic and maturational studies of aggressive behavior. In C. E. Clemente & D. B. Lindsley (Eds.), *Aggression and defense: Neural mechanisms and social patterns.* Vol. 5. *Brain function.* Berkeley: Univ. of California Press. Pp. 57–71.

Einsiedel, A. A., Jr. 1973. The development and modification of object preferences in domestic white leghorn chicks (*Gallus gallus*). Doctoral dissertation, Department of Psychology, Indiana University.

Fleener, D. E. 1967. Attachment formation in humans. (Doctoral dissertation, Indiana University.) Ann Arbor, Michigan: University Microfilms. No. 6872–12.

Fuller, J. L. 1967. Experiential deprivation and later behavior. *Science, 188,* 1645–1652.

Fuller, J. L., & Sjursen, F. H., Jr. 1967. Audiogenic seizures in eleven mouse strains. *Journal of Heredity, 58,* 135–140.

Garstang, W. 1922. The theory of recapitulation: A critical restatement of the biogenetic law. *Journal of the Linnean Society of London Zoology, 35,* 81.

Garvey, C., & Hogan, R. 1973. Social speech and social interaction: Egocentrism revisited. *Child Development, 44,* 562–568.

Geist, V. 1971. *Mountain sheep: A study in behavior and evolution.* Chicago: Univ. of Chicago Press.

Gottleib, G. 1970. A stranger in the land of the homologies. Paper presented at the Philosophy of Science Symposium, University of North Carolina, Chapel Hill, April 17–18, 1970.

Gottlieb, G. 1971. *Development of species identification in birds: An inquiry into the prenatal determinant of perception.* Chicago: Univ. of Chicago Press.

Harper, L. V. 1971. The young as a source of stimuli controlling caretaker behavior. *Developmental Psychology, 4,* 73–88.

King, M., & Wilson, A. C. 1975. Evolution at two levels in humans and chimpanzees. *Science, 188,* 107–116.

Kummer, H. 1971. *Primate societies.* Chicago: Aldine-Atherton.

Kuo, Z. Y. 1960. Studies on the basic factors in animal fighting: VII. Interspecies co-existence in mammals. *Journal of Genetic Psychology, 97,* 211–225.

Kuttner, R. 1960. An hypothesis on the evolution of intelligence. *Psychological Reports, 6,* 238–289.

Lagerspetz, K. M. J., & Lagerspetz, K. Y. H. 1971. Changes in the aggressiveness of mice resulting from selective breeding, learning, and social isolation. *Scandinavian Journal of Psychology, 12,* 241–248.

Lagerspetz, K. Y. H., Tirri, R., & Lagerspetz, K. M. J. 1968. Neurochemical and endocrinological studies of mice selectively bred for aggressiveness. *Scandinavian Journal of Psychology, 9,* 157–160.

Lehrman, D. S. 1970. Some semantic and conceptual issues in the nature–nurture problem. In L. R. Aronson, E. Tobach, D. S. Lehrman, & J. S. Rosenblatt (Eds.), *Development and evolution of behavior: Essays in memory of T. C. Schneirla.* San Francisco: Freeman. Pp. 17–52.

Lewis, M. 1971. Stability of attachment behavior: A transformational analysis. Paper presented at the meeting of the Society for Research in Child Development, Minneapolis, April, 1971.

Lorenz, K. 1966. *On aggression.* New York: Harcourt, Brace, & World.

Lorenz, K. Z. 1974. Analogy as a source of knowledge. *Science, 185,* 229–234.

Maccoby, E. E., & Feldman, S. S. 1972. Mother-attachment and stranger reactions in the third year of life. *Monographs of the Society for Research in Child Development, 37,* No. 1 (Serial No. 146).

Mason, W. A. 1968. Scope and potential of primate research. In J. H. Masserman (Ed.), *Science and psychoanalysis.* Vol. 12. *Animal and human.* New York: Grune & Stratton. Pp. 101–118.

Mason, W. A., & Kenny, M. D. 1974. Redirection of filial attachments in rhesus monkeys: Dogs as mother surrogates. *Science, 183,* 1209–1211.

Masters, J. C., & Wellman, H. M. 1974. The study of human infant attachment: A procedural critique. *Psychological Bulletin, 81,* 218–237.

McGrew, W. C. 1972. *An ethological study of children's behavior.* New York: Academic Press.

Monahan, L. C. 1975. Mother–infant and stranger–infant interaction: An ethological analysis. Doctoral dissertation, Department of Psychology, Indiana University.

Nowlis, V. 1941. Companionship preference and dominance in the social interaction of young chimpanzees. *Comparative Psychological Monographs, 17,* No. 1 (Whole No. 85).

Rheingold, H. L. (Ed.) 1963. *Maternal behavior in mammals.* New York: Wiley.

Rheingold, H. L., & Eckerman, C. O. 1973. Fear of the stranger: A critical examination. In H. W. Reese (Ed.), *Advances in child development and behavior.* Vol. 8. New York: Academic Press.

Salzen, E. A., & Meyer, C. C. 1967. Imprinting: Reversal of a preference established during the critical period. *Nature, 215,* 785–786.

Simmel, E. C. 1973. The analysis of the effects of early experiences. *Developmental Psychobiology, 6,* 510.

Stayton, D. J., & Ainsworth, M. D. S. 1973. Development of separation behavior in the first year of life: Protest, following, and greeting. *Developmental Psychology, 9,* 213–225.

Suomi, S. J., & Harlow, H. F. 1972. Social rehabilitation of isolate-reared monkeys. *Developmental Psychology, 6,* 487–496.

Thompson, D. W. 1942. *On growth and form.* Cambridge: Cambridge Univ. Press. (Originally published, 1917.)

Tinbergen, N. 1974. Ethology and stress diseases. *Science, 185,* 20–26.

Zajonc, R. B., Reimer, D. J., & Hausser, D. 1973. Imprinting and development of object preference in chicks by mere repeated exposure. *Journal of Comparative and Physiological Psychology, 83,* 434–440.

Chapter

7

CULTURAL EVOLUTION OF LEARNED LANGUAGE IN CHIMPANZEES[1]

Roger S. Fouts
University of Oklahoma

Joseph B. Couch
University of Oklahoma

Since Frank Beach's 1949 presidential address to the American Psychological Association, a cloud of doom has hung over the comparative psychologist's head. We seem in many ways like the last of a dying species. In spite of this precarious position, we shall initially move from what security we have in our empirical background into the rather tenuous world of theory and conjecture, although stepping out on a theoretical limb can be an uncomfortable position for a comparative psychologist. Theoreticians seem to be

[1] Some of the studies reported in this paper were supported by HEW grant 1R01 MH23511-01 EPR.

able to make profound statements and then, when presented with empirical data or new data, they are able to say quite easily, "Well, we didn't mean what we said earlier, and we take it all back." Our limb-resting position, then, will be more like that of the chimpanzee. In observing chimpanzees you will notice that when they venture out on a limb and stand on it they also have a firm grasp on another limb. Therefore, as we stand erect on the limb of theory we shall also hang on for dear life to the limb of empiricism.

In the past, the study of language and language development has not been allowed into the domain of comparative psychology. At this point it should be noted that the Gardners (1969, 1971) justifiably refer to research of American Sign Language acquisition in chimpanzees as *two-way communication* rather than *language*. It is a semantic way of avoiding a controversy, thus preserving an empirical background without moving too far into the field of theory. It appears as though a "behavioral apartheid" has developed in that language behavior is perceived by many as unique to human beings. If this were true, then there would be no basis for comparison, since one cannot compare a behavior across several species if that behavior exists only in one species. As a result, this area of inquiry has not been developed, and only rarely examined.

The attitude on the part of scientists in regard to language appears to have been influenced by several questionable assumptions concerning the nature of human beings, which have been accepted as fact for at least 2000 years. One such assumption is that human beings are unique. We have no argument with this. However, implicit in this statement is the assumption that *only* human beings are unique. With this we disagree, for the chimpanzee is also unique, as is the cockroach, the ant, the butterfly, and so on. Using the logic that uniqueness is confined to one species, one could arbitrarily designate that language shall be defined as that spoken by a certain individual. Language would then be confined to that individual, and no one else could be said to possess it. The pitfalls of using this type of logic are obvious, we feel.

Another questionable assumption stems from the first. That is, since human beings are unique, they are therefore not subject to the same biological and behavioral laws that other animals are. Human beings have consistently drawn a dichotomy between themselves and other animals, as a result of which they seem to assume that they are immune from or above the mechanisms that influence the biological and behavioral evolution of organisms. This implies a difference in kind rather than of degree between human beings and other animals. This philosophy contradicts Darwinian theory, which we view as positing a continuity between all organisms. It also violates an important basis underlying experimental and comparative psychology: There are laws and rules of behavior to which all organisms are subject. Nonhuman organisms are neither exempt from the principles and laws of biology nor those of learning and behavior simply because they are unique. Similarly, because human beings are unique does not mean that they lie outside of these principles and laws.

In reviewing the literature concerning human language and animal communication, one finds the definitions given for language emphasizing the differences between the two types of communication. The conclusion drawn, that animals are not capable of a language system like that of human beings, is based on a lack of evidence or negative evidence. That is, until recently there has been no evidence that any animal could use a human form of two-way communication. To draw the conclusion that animals are not capable of such a feat is comparable to saying, 200 years ago, "Human beings will never fly because they have never flown." Again, there is the persistence in forming a dichotomy between humans and animals.

We are proposing that language is a behavior, and that behavior exists along a continuum. Therefore, we also posit that there is a continuity of behavior between human beings and organisms similar to human beings. This continuity is what experimental and comparative psychology is based upon. It allows us, for example, to study rats instead of human beings in an effort to discover the general laws of learning.

Mayr (1972) points out that evolution has taken place for behavior as well as for physical characteristics. This notion is based upon the assumption that behavior has a biological basis, and therefore implies that a continuity exists.

Harvey Sarles (1972), an anthropologist at the University of

Minnesota, points out, "Biologically speaking, we expect continuity and relationship, not emergence and saltation. And we are rightfully suspicious when told that man is more than a bit outside of nature [p. 4]." In terms of language, Sarles goes on to state that there are two distinct theories of language origin: a sentence theory and a word theory. The sentence–grammar theory adheres to the position that human beings are unique, and therefore its proponents deny the value of comparative work. Sarles states that this theory holds to the position that sentences are ideas or thoughts and therefore are only in the human behavioral domain. Chomsky (1967), with his assumption that language is innately human, would fall into this category. Further, Chomsky's view that the appearance of language was due to a mutation within our species, pushing biological facts to an untenable position, has led Sarles to label him a Creationist's grammarian (i.e., human language was the result of divine creation). Word theorists, on the other hand, hold that animals sign and humans symbolize. As a result, the proponents of this theory maintain that the naming or labeling of objects is the paramount issue. With the appearance of Washoe's first signs, the word theorists' position was far less reasonable. Sarles notes that one theorist, J. Bronowski, retreated from the word theorist's position into the Creationist's grammarian position as a result of the work done by Gardner and Gardner (1969, 1971) and Premack (1971a) with chimpanzees.

In an article by Bronowski and Bellugi (1970) the check list approach to language is taken. This process involves the use of a predetermined list of the characteristics they feel language should contain. One then looks at the data and checks off on the list all the characteristics that were found. Their check list, small when compared to the expanding list of Hockett (1958) and Thorpe (1972), contains five characteristics: (1) delay between stimulus and utterance; (2) separation of affect from content; (3) prolongation of reference; (4) internalization; and (5) reconstitution. They concluded that Washoe had demonstrated all but the last characteristic. They defined reconstitution by stating, "Structure activity of reconstitution, which consists of two linked procedures—namely, a procedure of analysis, by which messages are not treated as inviolate wholes but are broken down into smaller parts, and a procedure of synthesis by which the parts are rearranged to form other messages [p. 670]." Based on this statement, the conclusion Bronowski and Bellugi came to is a shining example of emergence. They concluded that "What the example of Washoe shows in a profound way is that it is the process of total reconstitution which is the evolutionary hallmark of the human mind. And for which so far we have no evidence in the mind of the nonhuman primates even when he [Washoe is a she, by the way] is given the vocabulary ready made [p. 673]."

Rather than address Bronowski and Bellugi's statements directly, we will revert back into the world of empiricism and present data from a group of experiments dealing with language acquisition in the chimpanzee, leaving readers free to draw their own conclusions.

PROJECT WASHOE

Descriptions of Project Washoe have been so widely published that we shall set forth here only a brief summary.

Project Washoe was started in 1966 by Alan and Beatrice Gardner (1971). When Washoe arrived at the Gardners' home she was between 10 and 14 months of age. The Gardners, rather than adhering to the typical methodology used by experimental psychologists, raised Washoe in what would be considered by human standards an enriched environment. Washoe was immersed in an environment of American Sign Language for the Deaf. American Sign Language, or Ameslan, is a gestural language used by the deaf people in the United States and Canada. These gestures have specific movements that begin and end in a prescribed manner in relation to the signer's body. These movements or signs are analogous to words in a spoken language. This language has a bit more iconicisity than does vocal English, due probably to the fact that it is a visual mode mapping a visual world. There are, however, some signs that show no iconicisity and appear to be completely arbitrary.

Washoe employed several methods in acquiring signs. Certain signs already seemed to be in her repertoire of gestures. The *gimme* sign or begging gesture with arm extended is produced by many chimpanzees when they desire something, and so it was with Washoe. When excited she would shake her hand at the wrist, a gesture quite similar to the *hurry* sign in Ameslan. Training for these gestures consisted of simply shaping them into use in the appropriate context.

Most of Washoe's signs, however, were introduced through the technique of molding (Fouts, 1972). This process involves physically molding the chimpanzee's hand or hands into the proper position for a given sign in the presence of the exemplar for that sign. The molding procedure is then slowly faded away so that the chimpanzee is eventually making the sign by himself in the presence of the exemplar.

Washoe also picked up signs through observational learning, including several signs the Gardners did not want her to learn. One such sign was *smoke.* Several of the workers on the project smoked cigarettes and when one person would run out they would sign in front of Washoe to a fellow worker "*You have smoke* [cigarette]?" Washoe, of course, was quite interested in the procedure whereby people would put these objects in their mouths and then blow out smoke. Soon the *smoke* sign was observed in Washoe's vocabulary without any intentional training. Another example of observational learning involved the sign *toothbrush.* This sign is quite iconic in that it is made by rubbing the extended forefinger of one hand back and forth across the front of the teeth. Again, no active training on the part of the experimenters took place, but soon Washoe had added that sign also to her vocabulary.

The Gardners kept daily records of Washoe's signing. Vocabulary review sessions were held for her each day. Washoe's use of combinations of signs and the context in which they occurred were also recorded.

Testing of Washoe's vocabulary was conducted using blind procedures to control for cueing. The procedure the Gardners found most effective was the double-blind slide test. This test involved the presentation of transparencies of items in Washoe's vocabulary on a back-projecting screen. This screen was covered by a sliding door that Washoe could open. She initiated the test by pushing back the sliding door, which enabled her to view the transparency on the screen. Standing next to Washoe, but unable to see the screen, was an experimenter. After Washoe opened the door, this experimenter asked her, *"What's that?"* Washoe signed, and the response was recorded by the experimenter. Since this person did not see the screen, there was no possibility for cueing Washoe as to the correct sign. In order to obtain interobserver reliability, another experimenter recorded Washoe's responses from behind a one-way mirror above the screen. This person was also unable to see the screen, and therefore could not cue Washoe either. In short, the only primate able to see the screen and respond to it was Washoe. Using this procedure, Washoe signed 53 out of 99 items correctly. It is also interesting to note that on this test Washoe's errors fitted into meaningful categories. For example, a transparency of an animal had a high probability of being responded to with a sign for another animal. The same was true for objects in the food and grooming categories. Washoe's errors showed that she had the concepts for the categories of items in her vocabulary.

The Gardners also looked at rule-following behavior in Washoe's signing (they were reticent to use the word "syntax" because of its linguistic implications). Over a period of 2 months, they examined Washoe's use of the pronouns *you* and *me* with an action verb. Originally, Washoe preferred an order of *you–me*–action verb (*you* preceding *me* and the action verb over 90% of the time). However, after about 5 weeks she abruptly changed her preference in favor of the order *you*–action verb–*me*. Over this 2-month period *me* had preceded the action verb 60% of the time and followed it 40%. The reason for Washoe's change in preference of word order is still open to speculation.

EXPANDING HORIZONS IN SIGNING CHIMPANZEES

Booee, Bruno, Cindy, and Thelma

After the publication of Project Washoe, some people posited that Washoe was simply an exceptional chimpanzee with an abnormally high intelligence level. Her critics took the position that other chimpanzees would never be able to match the feats of this mutant intellectual genius. Therefore, in 1970, when Washoe and Dr. Fouts arrived in Norman, Oklahoma at the Institute for Primate Studies, one of the first tasks was to demonstrate the ability of other chimpanzees to acquire sign language. This experiment (Fouts, 1973) would not only help to determine whether other chimpanzees could acquire Ameslan, but would also investigate individual differences among chimpanzees in regard to ease or difficulty in acquiring sign language. The data from this experiment would also produce a comparison between signs; that is, are some signs easier to acquire than others?

Four chimpanzees were used as subjects in this experiment: Booee, Bruno, Cindy, and Thelma. Each subject was taught 10 signs utilizing molding as the training method. Acquisition time was measured in minutes, with a criterion for acquisition of five consecutive unprompted responses. There was no ceiling established for the amount of training, and it continued until the criterion was met. After each sign was acquired, it was integrated into the training sessions. This was done to make sure the chimpanzees were not making one sign to every exemplar. After all four chimpanzees acquired all 10 signs a double-blind box test was administered. (This procedure was quite similar to the double-blind slide test with the exception that the actual exemplars were placed in a box so that only the subject was able to see them.)

The results demonstrated individual variation in acquisition across signs. The range was 9.7 min to 316 min for acquiring a sign. The *look* sign was relatively slow to be acquired, whereas the signs for *listen* and *shoe* reached acquisition rapidly.

There were, as expected, individual differences across chimpanzees when comparing mean times to acquisition. Booee, for example, had a mean time of 54 min, while Cindy had a mean time of 79 min, and Bruno 159 min. These differences can be explained in terms of behavioral differences, or, if you prefer, personality differences in the chimpanzees. For example, Booee was the type of chimpanzee who was extremely food oriented. Just a bowl of raisins was a special delight to Booee and greatly increased his enthusiasm. Cindy, on the

Figure 7.1 Booee signs *baby* to appropriate exemplar.

other hand, needed what clinicians refer to as "unconditional positive regard." She demanded not only raisins but compliments and reassurance at every turn. Thelma was very easily distracted, which was reflected in her 136 min of mean time to acquisition. During training, a car would go by and she acted as though she had never seen such a strange thing. Or perhaps a fly might enter the training cage, and that too would be grounds for an immediate cessation of training.

Bruno was altogether different. With his 159-min mean, he became the first chimpanzee in danger of not acquiring signs. The experimenters decided to increase his incentive from raisins to apple slices. No change in behavior was noted. Then apple slices were abandoned in favor of banana slices. Again, no change in behavior. Finally, the ultimate in treats for a chimpanzee, a coke. At this point, Booee would have been disabled. But with Bruno there was no change. The situation was extremely frustrating. Bruno was attentive, he allowed his hands to be molded, but still—nothing. One day, after about five such sessions, an experimenter was trying to teach Bruno the *hat* sign. Bruno's behavior was the same, and finally the experimenter threatened Bruno. Immediately he began to sign *hat, hat, hat.* It is not surprising that Bruno was given the sign for proud as his name sign.

In testing, all the chimpanzees were above the chance level of responding, with Cindy the lowest at 26% correct. This behavioral decrement was not surprising. Cindy would usually sign the first two or three items in the test correctly, and then continue to use the last correct sign for every other item in the list. In the testing situation, praise was not administered, and Cindy responded accordingly. Thelma scored 58% correct, Booee 59% correct, and, going directly against his training behavior, Bruno scored 90% correct. It had been demonstrated that other chimpanzees could acquire signs in Ameslan.

Figure 7.2 Bruno makes the *hat* sign.

LUCY'S CONCEPTS OF THE WORLD OF FRUITS AND VEGETABLES

Lucy is a chimpanzee who has been reared in species-isolation in a human home since she was 2 days old. She is now 7 years old. At the time of the experiment she had been trained in Ameslan for over 2 years and had a vocabulary of 75 signs.

This experiment (Mellgren, Fouts, & Lemmon, 1973) was conducted in order to examine the relationship between generic and specific signs. It also tested the conceptual ability of a chimpanzee in regard to the category of items in its vocabulary.

Lucy had five food-related signs in her vocabulary: *food, fruit,* and *drink,* which she used in a generic manner; and *candy* and *banana,* which she used in a specific manner. The study was designed to teach Lucy a new sign and then record whether or not it became specific or generic in relation to a category of items. The sign chosen was *berry,* with a cherry serving as the exemplar. The category of items consisted of 24 different fruits and vegetables ranging from one-quarter of a piece of watermelon and a grapefruit to small berries and berry-like items such as blueberries, cherry tomatoes, and radishes.

Lucy was presented with each of these 24 fruits and vegetables and asked *"What's that?"* in Ameslan. (These items were interspersed between at least two items in her vocabulary that were not among the 24 fruit and vegetable items.) She was allowed to pick the item up, play with it, chew it, or do whatever she wanted to with it. These items were presented in a different random order for each day of data collection and her responses were recorded.

In order to determine Lucy's usual responses to the 24 fruit and vegetable items, 4 days of baseline data were collected prior to the introduction of the *berry* sign. After the fourth day she was taught the *berry* sign, as mentioned before, with a cherry as the exemplar. During the next 4 days the *berry* sign remained specific to cherries. After the eighth day Lucy was again taught the sign for *berry,* but this time a blueberry was used as the exemplar. For the next 2 days she called the blueberries *berry* and then went back to her original sign for *blueberry.* However, she continued to sign *berry* to the

Figure 7.3 Lucy signs *fruit* to Dr. Fouts during one of her training sessions. [Photo courtesy of Dr. Roger L. Mellgren, Psychology Department, University of Oklahoma.]

cherries. Lucy obviously preferred to use the *berry* sign in a specific sense.

The experiment also examined Lucy's conceptualizations of fruits and vegetables. It was determined that Lucy dichotomized the two categories in her signing. The *fruit* sign was used 85% of the time to label fruit items as opposed to 15% for the *food* sign. For vegetables the reverse was true. Lucy preferred the *food* sign (65%) for food items as opposed to the *fruit* sign (35%). It is also interesting to note that the *smell* sign was used 65% of the time to describe the four citrus fruits. She referred to these as *smell fruits* probably due to the odoriferous qualities experienced when biting into the skin of a citrus fruit.

Lucy also constructed several novel combinations in an attempt to describe some of these 24 items. When presented with a radish, Lucy signed *fruit food* or *drink* for the first 3 days. On the fourth day she bit into the radish, spit it out, and signed *cry hurt food.* From that point on she used either *cry* or *hurt* to describe the radish. In describing a watermelon, Lucy signed *candy drink* or *drink fruit* even though the experimenter used the signs *water* and *melon.* These novel combinations not only demonstrated Lucy's ability to form new combinations from the signs in her vocabulary, but also revealed Lucy's ability to express her concepts of items in her environment.

ALLY: THE SIGN AND THE SOUND

As one might imagine, the home-reared chimpanzees have had a great deal of exposure to vocal English. The human parents of these chimpanzees have often stated that they believed their chimpanzee child could understand a certain amount of spoken English. Therefore, a study was conducted to determine the relationship between their vocal English understanding vocabularies and their Ameslan vocabularies.

The experiment (Fouts, Chown, & Goodin, 1973) involved a young male chimpanzee, Ally, who was being home-reared at the time. His understanding of 10 English words was tested in a pretraining procedure. Ally was given vocal commands such as: "Give me the spoon," "Pick up the spoon," "Find the spoon," etc. Ally had to obey the command by choosing the requested object from a group of several other objects. The criterion set was obeying a command five consecutive times. When the criterion was met, Ally was considered to have an understanding of these 10 vocal English words.

In testing, these 10 English words were divided into two groups of 5 words each. The first experimenter then attempted to teach Ally a sign using only the vocal English word as the exemplar. A second experimenter then tested Ally on all five of the objects corresponding to the vocal English words without knowing which, if any, had been either taught or acquired. The reverse procedure was used for the second list of 5 words. The results demonstrated that Ally was able to transfer the sign taught for the vocal English word to the object representing that word. This process is similar to second language acquisition in humans and also has cross-modal implications.

Comprehension of Novel Commands

As mentioned earlier, it has been demonstrated that chimpanzees are able to create novel sign combinations from familiar signs in their vocabulary. This next study (Chown, 1974) explored the ability of a chimpanzee to comprehend novel commands. Ally was also the subject of this experiment.

Ally was first taught to play a game of picking, on command, one of five objects out of a box and putting it in one of three locations. An example of such a command is: *"Put baby in purse."* Once Ally had caught on to the game, new items were put in the box and a novel location in which to place these items was added to the locations used in training.

In testing, the experimenter, who gave Ally the commands, placed the five new items in a box into which he could not see. A screen was positioned between the experimenter and the three possible locations in which the item could be placed. This procedure controlled for any possible cueing of Ally by the experimenter.

With five items to choose from and three possible locations in which to place items, chance level of responding was determined to be 1 out of 15 (approximately 7%). The results of the experiment demonstrated that for total commands Ally scored 40% correct. It should be noted that Ally is a very hyperactive chimpanzee, and he often grabbed the correct object then ran behind the screen before the experimenter finished signing the command. When broken down, the data revealed as much as a 60% difference between percentage correct for the object and percentage correct for the location. This means that the 40% figure for total commands probably is conservative in assessing Ally's comprehension of novel commands.

Productive Use of a Grammatical System

As noted earlier, some authors assumed that because Washoe had not evidenced grammar in her spontaneous combinations, the acquisition of such a system was beyond the capacity of a chimpanzee. However, later studies provided evidence of syntactic competence in the chimpanzee. Premack (1971a,b; Premack & Premack, 1972) demonstrated that a chimpanzee can be trained to use specific symbol orders to request fruit or other foods from a human trainer. Rumbaugh, Gill, and von Glaserfeld (1973) have shown that a chimpanzee can distinguish valid sentence beginnings from invalid ones. The previously mentioned Chown (1974) study demonstrated that a chimpanzee can comprehend novel commands generated by vocabulary substitution within familiar fixed command frames.

This study (Chown, Kimball, Couch, & Fouts, in preparation) was designed to demonstrate the active application of a syntactic system to express spatial relations between objects in the environment. Spatial relations between objects were easily specified by short sequences of signs, yet syntax was essential for expression of proper meaning. This relation was expressed by the order of the signs (subject, preposition, location) and the rules of the grammar generated all permissible sequences and no ungrammatical sequences. The system could not legitimately be called formally complex, but it did incorporate all of the essential aspects of language in rudimentary form.

Ally, then 5 years old, again served as the subject. He was taught the relations—"on," "in," and "under"—by setting physical objects (noun exemplars) in appropriate arrangements. He was then taught to describe the situation accurately in Ameslan. Tests of acquisition were conducted by asking Ally to describe situations. For example, a flower would be placed on a pillow, and then the experimenter would sign to Ally *"Where flower?"* In order to be correct, Ally then had to make the appropriate signs in the proper order. These testing situations varied from familiar (those used in training) to completely novel. Once again a double-blind procedure was used during testing to control for the cueing of Ally by an experimenter.

Figure 7.4 (a–c) Ally describes the appropriate relationship (ball under box) to researcher George Kimbal at the Institute for Primate Studies.

During testing Ally expressed the prepositions correctly 84.9% of the time in familiar situations (F) and 76.9% in novel situations (N). He correctly expressed the locations 77.9% (F) and 64.3% (N). The total relationship was expressed correctly 67.4% (F) and 50.0% (N). Although these figures were far above a chance level of responding, they were greatly affected by several interesting types of errors committed by Ally. When the novel location "purse" was used, a great many errors occurred. It was quite probable that this specific decrement was due to the physical attributes of the exemplar itself. An old red laundry bag with a white drawstring was used as the exemplar for "purse." When the exemplar was on the floor it was impossible for Ally to determine if the subject was "in" or "under" the purse since either condition resulted in a lump in the red bag. Furthermore, there was a conceptual confusion between this exemplar and the exemplar for "blanket." When laid on the floor, the red bag closely resembled a blanket, and Ally in fact made several such conceptual errors in responding.

On several occasions Ally would express a relationship such as *ball under box* as *ball under in box*. The latter response was always recorded as incorrect due to the use of two prepositions. However, later when asked *"What this?"* (pointing to the box), Ally often signed *in box*. Thus, it was possible that this error was a functional one insofar as the box was indeed used to carry the exemplars from the storage area to the training area. That is, the box was an object that items were placed "in," hence it became an *in box*. If the errors involving the purse and the box were removed, the entire relationship was expressed correctly 68.6% of the time.

It should be noted that Ally often did not sign the subject of the relationship. This was not counted as an incorrect response. However, if he did sign the subject it had to be correct or the entire relationship expressed was recorded as incorrect. Out of 240 trials he signed the entire relationship (subject, preposition, location) 44 times. The subject was signed correctly 42 times for 95.5% accuracy.

We feel that it was significant that in testing Ally was required to construct the relationships entirely on his own. Never did he make a grammatical error. The relationships expressed were always subject, preposition, location, or ____, preposition, location. Ally has demonstrated the ability to express a change in meaning through specific structural recombinations, specified by a set of grammatical rules. This indicates that a chimpanzee is capable of utilizing an active linguistic process.

LAGNIAPPE

As the subtitle implies, this next experiment (Beatty, Fouts, & McDivit, in preparation) brought more than the experimenters expected. John Beatty, an anthropologist from Brooklyn College, came to the Institute to examine the ability of a chimpanzee to differentiate between various chimpanzee calls. Beatty brought with him recordings of chimpanzees made by Peter Marler at the Gombe Stream Reserve. Washoe was chosen as the subject for this experiment, but it was soon apparent that she was more interested in Beatty than in participating in the experiment. Therefore, Bruno (approximately 8 years old at the time) was selected as the new subject for the experiment.

The first step was to teach Bruno the signs for *same* and *different*. This was accomplished by first presenting Bruno with two round red beads and teaching him the *same* sign. Then he was shown a red bead and a blue bead and taught the sign *different*. Then two identical dolls were presented and Bruno was taught the *same* sign in their presence. This procedure continued until Bruno was responding reliably to objects with the signs *same* and *different*. Bruno was then exposed to tape recordings of both high- and low-pitched sounds, and again taught the *same* and *different* signs to the appropriate pairings of sounds. Once reliability was established for auditory stimuli, testing was begun.

Figure 7.5 Bruno signs *same* describing two identical red beads.

The following pairs of calls were presented to Bruno in testing: Chimpanzee A bark, Chimpanzee B scream; A bark, A scream; A bark, B bark; A bark #1, A bark #2; A bark #1, rerecording of A bark #1. Trial 1 on the first day produced a response of *different* by Bruno to each pair of calls. Then, on each of Trials 2–5 of the first day Bruno responded exactly the same way: *different; same, different; same, different; same; same.* Obviously a pattern had developed, and so the test was repeated the next day. Again (with the exception of the response to the fourth pair on Trial 3) Bruno's responses were exactly like those of Trials 2–5 on Day 1. It was thought at first that Bruno had failed the discrimination test by his signings of *same, different* to two pairs of calls. Further analysis, however, produced a most interesting result. Bruno was not only differentiating between calls, he was differentiating between the chimpanzees giving the calls. This makes sense, for certainly it is important for a chimpanzee to know not only what call is being given, but who is giving it. His responses were appropriate. For example, A bark, B scream is different both in chimpanzee and type of call *(different).* A bark, A scream is same chimpanzee, different call *(same, different).* A bark, B bark is also a *same, different* situation in that the type of call is the same but the chimpanzee is different. The last two pairs were interpreted by Bruno as same chimpanzee, same type of call *(same).* Bruno gave the experimenters much more than they had first imagined.

Proper methodology is basic to psychology, and so it should be. However, this experiment demonstrated quite well a point made by Wolfgang Kohler (1971) many years ago. If one overstructures the situation in terms of examining the behavior of organisms, the only behavior revealed in the organism is the behavior the experimenter is capable of preconceiving.

CLOSING REMARKS

With the exception of Project Washoe, the studies outlined in this chapter reflect the research involving ASL-responding chimpanzees already completed at the Institute for Primate Studies. Current research has expanded to include new areas of investigation. Studies designed to explore verbal discrimination processes in signing chimpanzees are underway, as are experiments in chimpanzee-to-chimpanzee ASL communication. As each new experiment is concluded and the data analyzed, we become more aware of how much information there is left to be acquired from the chimpanzee. The nature and the quality of the information we shall gather in the future seems at this point bound only by the limitations of our experimental imagination and creativity.

REFERENCES

Beatty, J., Fouts, R. S., & McDivit, A. In preparation. Same–different discrimination of chimpanzee vocalizations by an ASL responding chimpanzee. *Science.*

Bronowski, J., & Bellugi, U. 1970. Language, name, and concept. *Science, 168,* 669–673.

Chomsky, N. 1967. The formal nature of language. In E. Lenneberg, (Ed.) *Biological foundations of language.* New York: Wiley. Pp. 397–442.

Chown, W. B. 1974. Productive competence in a chimpanzee's comprehension of commands. Master's thesis, Psychology Department, University of Oklahoma.

Chown, W. B., Kimball, G. H., Couch, J. B., & Fouts, R. S. In preparation. Productive use of a grammatical system by a chimpanzee.

Fouts, R. S. 1972. The use of guidance in teaching sign language to a chimpanzee. *Journal of Comparative and Physiological Psychology, 80,* 515–522.

Fouts, R. S. 1973. Acquisition and testing of gestural signs in four young chimpanzees. *Science, 180,* 978–980.

Fouts, R. S., Chown, W. B., & Goodin, L. 1973. The use of vocal English to teach American Sign Language (ASL) to a chimpanzee: Translations from English to ASL. Paper presented at the Southwestern Psychological Association Meeting, Dallas, Texas, April 26–28, 1973.

Gardner, R. A., & Gardner, B. T. 1969. Teaching sign language to a chimpanzee. *Science, 165,* 664–672.

Gardner, B. T., & Gardner, R. A. 1971. Two way communication with an infant chimpanzee. In a Schrier & F. Stollnitz (Eds.), *Behavior of nonhuman primates.* New York: Academic Press. Pp. 117–183.

Hockett, C. F. 1958. *A course in modern linguistics.* Toronto: MacMillan.

Kohler, W. 1971. Methods of psychological research with apes. In M. Henle (Ed.), *The selected papers of Wolfgang Kohler.* New York: Liveright. Pp. 197–223.

Mayr, E. 1972. The nature of the Darwinian revolution. *Science, 176,* 981–989.

Mellgren, R. L., Fouts, R. S., & Lemmon, W. B. 1973. American Sign Language in the chimpanzee: Semantic and conceptual functions of signs. Paper presented at the Midwestern Psychological Association Meeting, Chicago, Illinois, May 3–5, 1973.

Premack, D. 1971a. Language in chimpanzees? *Science, 172,* 808–822.

Premack, D. 1971b. On the assessment of language competence and the chimpanzee. In A. Schrier & F. Stollnitz (Eds.), *Behavior of nonhuman primates.* New York: Academic Press. Pp. 186–228.

Premack, J., & Premack, D. 1972. Teaching language to an ape. *Scientific American, 227,* 92–99.

Rumbaugh, D., Gill, T. V., & von Glaserfeld, E. C. 1973. Reading and sentence completion by a chimpanzee (Pan). *Science, 182,* 731–733.

Sarles, H. B. 1972. The search for comparative variables in human speech. A symposium paper presented at the Animal Behavior Society Meetings, Reno, Nevada, June 13–16, 1972.

Thorpe, W. H. 1972. Comparison of vocal communication in animals and man. In R. A. Hinde (Ed.), *Non-verbal communication.* Cambridge: Cambridge Univ. Press. Pp. 27–47.

Postscript

SOCIAL ORGANIZATION AND THE EVOLUTION OF COMMUNICATION

MARTIN E. HAHN
William Paterson College

EDWARD C. SIMMEL
Miami University

The opportunity to edit a volume on communication and evolution was an enviable one. We began our project with two areas of mutual interest, selected some issues within these areas on which to concentrate, and chose nine contributors whose views might aid our own and our readers' thinking in the areas. If we have been successful in our venture, new ideas and some synthesis will be apparent.

It would be impossible to draw conclusions from the preceding chapters for all readers, since each reader brings unique experiences and interests—to say nothing of genotype. Consequently, in the final chapter, we will discuss a few issues in general and leave the reader with some interesting questions to ponder.

SELECTED ISSUES

MAN AND ANIMALS

At the risk of offending those to whom the issue is apparent, we re-emphasize the data and thoughts of Fouts and Couch, and Pitcairn and Eibl-Eibesfeldt: Man fits within the continuum of animal life—not outside it. The ability of Fouts' chimpanzees to manipulate concepts and the extraordinary ability of Bruno to outwit his experimenters by responding to multiple dimensions of a vocalization speak for themselves.

Of equal or greater importance, man and all other species are unique as a result of evolving under different circumstances. Though the uniqueness may provide the comparative psychologist with insurmountable obstacles in the pursuit of general explanations of behavior, intriguing researches are possible to discover the nature of the uniqueness. A recent example is contained in an article by King and Wilson (1975) in which they compare man and chimpanzee, discovering very few differences in protein structure. In explaining their findings, they conclude that the important anatomical and behavioral deviations between these species resulted not from structural gene differences but from regulator genes that affect the timing of protein synthesis. An analysis of the differential function of such regulator genes in the determination of the important differences between species would illustrate a great deal about the process of evolution.

COMMUNICATION AND EVOLUTION

In this volume, Klopfer argues that a behavioral process, particularly in birds and mammals, is extremely complex—the complexity results from constant feedback during development under quite variable circumstances. He concludes, "The pathways of causality and control are numerous . . . with many crossovers, so it is rarely if ever possible to isolate single heritable factors whose effects are inevitable and irreversible." This position seems to jeopardize any further discussion of the evolution of communication, even though Klopfer mentions that social behaviors (bill-tip-pecking in gull chicks) provide some increase in genetic fitness.

In view of Klopfer's notions, what are the relationships between genes, evolution, and communication? Though the answer to this question cannot be fully expressed in the current volume, Fuller has made a beginning in writing that the genetic contribution to some communicative behavior (coordinated patterns between organisms) is large for some behaviors and small for others; e.g., alarm calls probably have a strong genetic component, while communications for individual recognition are much subject to learning.

Genetic studies with dogs, birds, and crickets demonstrate strain or species differences, and hybrid animals that are variously intermediate between parent types indicate various types of genetic control. While we choose to leave a resolution between the positions of Fuller and Klopfer to our readers, we suggest that nothing more than semantic juggling is required.

The relationship between evolution and communication appears to be more complex and difficult to resolve. There has been no convincing evolutionary explanation for the synergistic capacities of sender and receiver. The results of Bentley and Hoy cited by Fuller indicate that, in crickets, genes that establish a neurologic structure allowing the male to transmit song pattern also establish a similar neurologic structure in the female that allows reception of song.

Other explanations could include the evolution of wide-band receivers and transmitters that would favor the possibility of communications with other individuals who have similar, but not identical, mutations. As Fuller indicates, research opportunities in this area abound.

Communication and Social Structure

Most social behaviors are the prerequisites of more complex ones. Sensory contact is necessary for the occurrence of attraction–

repulsion, which is in turn necessary for aggregation, attachment, or communication, and so on. Communication is particularly central to the web of social relationships that comprise social organizations. The flow diagrams of Pitcairn and Eibl-Eibesfeldt attest to this centrality. In wolves, birds, dogs, crickets, mice, and primates, the role of communication is to establish and maintain a social organization with a minimum of aggression. In our own research with the house mouse, for example, postural and vocal communications contribute to the establishment and maintenance of dominance between males. Defeated males exhibit a defense posture and subsonic squeaking that are effective in terminating attack by other males.

The precise contributions made by other types of communication in mouse social structure are not as lucid. Olfactory communications allow individual recognition, which may contribute to dominance and does contribute to mating behavior and perhaps to genetic variability in the population via the Bruce effect.

A variety of approaches for the further analysis of the role of communication in social organization exists in the preceding chapters. We mention three of great interest.

Sequence analysis in social groups provides a unique vehicle for establishing cause and effect relationships among behaviors. The effect of a smile or vertical tail wagging on a receiver can be assessed by this technique.

Genetic studies of the composition of variation in communication patterns provide information about their relative malleability to the environment. More important, such studies analyze the means by which genetic and experiential information interact in the formation of a behavior.

Very interesting new data should follow the study of interactions by chimpanzees who communicate in sign language. If sign is used for communication, the experimenters may observe the establishment of a culture that differs markedly from that existent in chimpanzees not using sign.

REMAINING QUESTIONS

We would like to leave our readers the following questions and statements that are based on the preceding chapters and have great merit for continued discussion and research.

1. The concept of species-specific behavior, in particular species-specific communication, is highly questionable in light of Scott's point that the concept is grounded in typological thinking and is not consistent with current thought on within-species variation and evolution. Perhaps the concept should be discarded.

2. The concepts of homology and analogy have been generally used to discuss the evolution of behavior. Klopfer indicates the difficulties with the use of these concepts, particularly as behavior is reduced to the biochemical level. Is the use of homology and analogy still justified on heuristic grounds?

3. Fuller states that it is more efficient in the long run for the environment to shape the gene pool than to start anew in each generation to fit organisms to available niches. Cairns proposes that selection would favor plasticity for the development of social behavior. Is it possible that life span determines the control of social behavior such that short-life-span animals exhibit less intraspecific variation in social behavior, while long-life-span animals have evolved the "luxury" of less stereotyped means of social interaction?

REFERENCE

King, M. C., & Wilson, A. C. 1975. Evolution at two levels in humans and chimpanzees. *Science, 188,* 107–116.

AUTHOR INDEX

Numbers in italics refer to the pages on which the complete references are listed.

L

M

N

P

R

S

T

V

SUBJECT INDEX

L

M

N

O

P

R

S

A 6
B 7
C 8
D 9
E 0
F 1
G 2
H 3
I 4
J 5